A scholarly, and profoundly interdisciplinary, approach to stem cell research that stands the conventional model of scientific development on its head. Far from offering only a secondary and peripheral commentary on more technical matters, questions of ethics, epistemology and socio-technical process are given proper significance. Cells and clinical practices, philosophy and social science flow together in this fascinating and very timely collection. As the editors convincingly argue, the social sciences and humanities do not simply study stem cell research. They also help configure it.

Alan Irwin, Professor in the Department of
Organization, Copenhagen Business School, Denmark

Reducing stem cell research to issues of the moral status of embryos is not grasping its dimensions. This excellent interdisciplinary book highlights the central roles epistemic, social, ethical and political factors play in its formation and recent developments. An essential work for all who want to understand this life science in its societal complexity.

Ilhan Ilkiliç, Professor of Health Sciences, Istanbul
University, and Member of the German Ethics Council

I think it is a most interesting book, providing new, innovative and very important perspectives on stem cell research/science. It provides a panoramic account of the forces at play in its constitution: disciplines, techniques, specialties, commercial interests, medical concerns, ethical and political factors and more. It does so stressing complementarities and diversity in the dynamics of this wide variety of factors, without privileging one set of factors over another. A non-reductionist, comprehensive understanding of the development of contemporary science, and medical science in particular, requires such a broad grasp.

Rob Hagendijk, Professor of STS, University of
Amsterdam, The Netherlands

The Matrix of Stem Cell Research

Stem cell research has been a problematic endeavour. For the past twenty years it has attracted moral controversies in both the public and the professional sphere. The research involves not only laboratories, clinics, and people but ethics, industries, jurisprudence, and markets. Today it contributes to the development of new therapies and affects increasingly many social arenas. The matrix approach introduced in this book offers a new understanding of this science in its relation to society. The contributions are multidisciplinary and intersectional, illustrating how agency and influence between science and society go both ways.

Conceptually, this volume presents a situated and reflexive approach for philosophy and sociology of the life sciences. The practices that are part of stem cell research are dispersed, and the concepts that tie them together are tenuous; there are persistent problems with the validation of findings, and the ontology of the stem cell is elusive. The array of applications shapes a growing bioeconomy that is dependent on patient donations of tissues and embryos, consumers, and industrial support. In this volume it is argued that this research now denotes not a specific field but a flexible web of intersecting practices, discourses, and agencies. To capture significant parts of this complex reality, this book presents recent findings from researchers, who have studied in-depth aspects of this matrix of stem cell research.

This volume presents state-of-the-art examinations from senior and junior scholars in disciplines from humanities and laboratory research to various social sciences, highlighting particular normative and epistemological intersections. The book will appeal to scholars as well as wider audiences interested in developments in life science and society interactions. The novel matrix approach and the accessible case studies make this an excellent resource for science and society courses.

Christine Hauskeller is Professor of Philosophy, teaching bioethics, social and feminist philosophy at the University of Exeter, UK. She conducts empirical philosophical studies on the processes of knowledge production in the life sciences and their intersections with different forms of valuation and normativity. Prof. Hauskeller has published widely on stem cell research for the past twenty years and has led many projects on its developments and translation. Appointments to advisory and governance boards include the UK Biotechnology and Biological Sciences Research Council's Biosciences for Society Strategy Advisory Panel and the German Federal Government's Central Ethics Committee Stem Cell Research.

Arne Manzeschke is Professor of Anthropology and Ethics at the Lutheran University of Applied Sciences Nuremberg, Germany; Director of the Institute of Ethics and Anthropology in Health Care; President of Societas Ethica, European Society for Research in Ethics; and vice chairman of the Bavarian Ethics Commission on Preimplantation Genetic Diagnosis. Arne was a member of the Bavarian Research Network Induced Pluripotent Stem Cells (ForIPS). He conducts ethical research in the field of bioethics and the ethics of technology, especially digitalization and the human-machine interaction.

Anja Pichl is a research fellow at the interdisciplinary working group *Gentechnology Report* at the Berlin-Brandenburg Academy of Sciences (BBAW) and member of the research training group 2073 *Integrating Ethics and Epistemology of Scientific Research* at Bielefeld University, funded by the Deutsche Forschungsgesellschaft (DFG). Together with Arne Manzeschke, she organized the international summer school *Pluripotent stem cells: scientific practice, ethical, legal, and social commentary* and was a member of the Bavarian Research Network Induced Pluripotent Stem Cells (ForIPS).

Genetics and Society
Series Editors: *Ruth Chadwick, former Director of Cesagene,*
Cardiff University
John Dupré, Director of Egenis, Exeter University
David Wield, Director of Innogen, Edinburgh University
Steve Yearley, former Director of the Genomics Forum,
Edinburgh University.

The books in this series, all based on original research, explore the social, economic and ethical consequences of the new genetic sciences. The series is based in the Cesagene, one of the centres forming the ESRC's Genomics Network (EGN), the largest UK investment in social-science research on the implications of these innovations. With a mix of research monographs, edited collections, textbooks and a major new handbook, the series is a valuable contribution to the social analysis of developing and emergent bio-technologies.

Knowing New Biotechnologies
Social Aspects of Technological Convergence
Edited by Matthias Wienroth and Eugenia Rodrigues

Cybergenetics
Health genetics and new media
Susan Kelly, Anna Harris, Sally Wyatt

Psychiatric Genetics
From Hereditary Madness to Big Biology
Michael Arribas-Ayllon, Andrew Bartlett and Jamie Lewis

The Matrix of Stem Cell Research
An Approach to Rethinking Science in Society
Edited by Christine Hauskeller, Arne Manzeschke, and Anja Pichl

For more information about this series, please visit https://www.routledge.com/Genetics-and-Society/book-series/GANDS.

The Matrix of Stem Cell Research

An Approach to Rethinking Science in Society

Edited by
Christine Hauskeller,
Arne Manzeschke, and
Anja Pichl

Federal Ministry
of Education
and Research

LONDON AND NEW YORK

First published 2020
by Routledge
2 Park Square, Milton Park, Abingdon, Oxon OX14 4RN

and by Routledge
605 Third Avenue, New York, NY 10017

First issued in paperback 2020

Routledge is an imprint of the Taylor & Francis Group, an informa business

British Library Cataloguing-in-Publication Data
A catalogue record for this book is available from the British Library

Library of Congress Cataloging-in-Publication Data
Names: Hauskeller, Christine, 1964– editor.
Title: The matrix of stem cell research : an approach to rethinking science in society / edited by Christine Hauskeller, Arne Manzeschke, and Anja Pichl.
Description: Milton Park, Abingdon, Oxon ; New York, NY : Routledge, 2020. | Includes bibliographical references.
Identifiers: LCCN 2019017100 | ISBN 9781138098527 (hbk) | ISBN 9781315104386 (ebk)
Subjects: LCSH: Stem cells—Research—Moral and ethical aspects.
Classification: LCC QH588.S83 M38 2020 | DDC 174.2/8—dc23
LC record available at https://lccn.loc.gov/2019017100

ISBN 13: 978-0-367-72683-6 (pbk)
ISBN 13: 978-1-138-09852-7 (hbk)
ISBN 13: 978-1-315-10438-6 (ebk)

Typeset in Times New Roman
by codeMantra

In memoriam
   Wolfgang Bender (25 March 1931 – 15 July 2017)
   The teacher and friend who motivated this research.
   Christine

Contents

List of figures

List of tables

Acknowledgements

This matrix approach, as conveyed through the format of this book, evolved from the editors' continuing discussions following the international summer school 'Pluripotent stem cells: scientific practice, ethical, legal, and social commentary'. Several of the contributors to this volume presented at the event, whereas others were invited to contribute later, and we are grateful to all the authors for their willingness to provide original articles on their special research projects in the stem cell matrix.

We also express our gratitude to the institutions who supported us through this process, among them first and foremost the German Federal Ministry of Education and Research (BMBF) who awarded Arne Manzeschke generous funding for organizing the summer school in Bavaria and the project's public closing event in April 2017 as well as for the edition of this volume in English (funding code FKZ: 01GP1482). The Bavarian research network on human induced pluripotent stem cells, ForIPS, funded by the Bavarian State Ministry of Science and the Arts (funding code D2-F2412.26), in which Arne and Anja were active between 2014 and 2017, also supported the project, and so did the University of Exeter, with top-up funding for copy-editing. We further thank the conscientious peer reviewers who supported the development of the chapters, and, last but not least, we endorse and thank Dr Peter Hughes for his thoughtful editing, close attention to detail, and timely delivery of copy-edited texts – whilst concurrently completing his philosophy PhD.

1 Knowledge and normativity

A matrix of disciplines and practices

Christine Hauskeller, Anja Pichl, and Arne Manzeschke

Abstract

This chapter introduces the matrix approach as a method of studying the life sciences. It builds on insights from the philosophy and sociology of science over the past decades against which it is laid out. The contribution of this new approach is that it reconfigures the life sciences as multidisciplinary and multi-institutional societal projects. The case study of stem cell research shows that one cannot separate internal and external, that is epistemic, social, ethical, and political factors affecting the development of a life science. The matrix approach allows for the examination of this complex and dynamic web of interests and societal practices and highlights the important role of legal, normative, technical, and political conditions and activities in making the science work in specific ways. Life sciences operate in a matrix of diverse societal intersections and spheres of dominance. The approach especially reframes the role and scope of the humanities and social sciences. It prepares the ground for a self-critical reflection in bioethics, social sciences, and philosophy. These disciplines do not just analyse stem cell research as an object, from a distance, but co-configure it and shape its contemporary gestalt and practices.

Introduction

The matrix approach is a new method for investigating the developing life sciences in the context of societal demands and practices. It configures the social practices and institutions that engage with the life sciences as constitutive factors and parts of said development. The matrix approach builds on theories in the philosophy, history, and sociology of science, and aims to enhance critical self-reflexivity in the life sciences as well as in the social sciences and humanities. The case study is stem cell research, which was developed from the early 2000s and is entangled with more and more social and institutional contexts as well as academic arenas.

After some situating remarks, we begin to discuss the matrix approach by first distinguishing it from the conventional narrations of the breakthrough myth in science journals and institutions. We then discuss how the history of science has increasingly included socio-political and normative dimensions into its analyses of how science advances, changes, and the reasons for this – pointing out that the normative ethical dimension is still being under-represented in concrete science studies. We then question whether stem cell science can be understood as a unified field due to the obscure object that lends it its name, or due to the aim of advancing new clinical treatments that is supposedly shared by the partaking life science disciplines. The literature has shown it to be the case that, and why, these perspectives are falling short, and the matrix approach offers the alternative of a situated approach to life sciences in and as social practice. The chapter closes with a brief overview of the segments in the stem cell matrix which the individual book chapters bring out.

The matrix approach

Our method is based on problematization. Problematization means putting the issue in problem form and identifying it as a problem for politics (cf. Foucault, 1997, 114). It follows Paul Rabinow, who suggested not to take descriptions of a current 'situation as given' (cf. Rabinow, 2003, 15ff.), referring to Michel Foucault's suggestion to examine discourses using multiple methods and perspectives. Critical normative interactions with stem cell research, undertaken in philosophical epistemology and ethics as well as in empirical Science and Technology Studies (STS), ask and give accounts of both the problematizations that inform which 'facts' about a science are presumed as given and not spelled out or questioned, as well as focussing the analysis on the power relations in and around stem cell research.

Terminologically, we chose the matrix term not least as a nod to Thomas Kuhn's 'disciplinary matrix', defined in the postscript to *The Structure of Scientific Revolutions* as: 'the entire constellation of beliefs, values, techniques, and so on shared by the members of a given community' (Kuhn, 2012, 174). Like Kuhn, we explore the constitutional conditions for the development, contents, and constellations of science. But we move beyond looking at science as if it developed largely 'insulated' from society (Kuhn, 2012, 163). We widen the term to include analyses of the power relations, with the socio-economic, political, cultural, and epistemic dynamics and forces in stem cell research discourses and practices.

Political governance, funding policies, and organizations of research are comparatively transparent societal conditions shaping science. Furthermore, public funding arguably entails a responsibility put upon science with the aim of benefitting the public good (e.g. European Commission, 2009). The social, economic, and intellectual conditions of science are, we propose, not additions on top of the 'real' intra-scientific developments. They are an integral part of scientific development. In stem cell research this is particularly readily apparent, given the public debates about the ethical acceptability of hESC (human embryonic stem cell) research still in force in many countries, and the legal and regulatory limitations established to restrict conduct of stem cell clinical trials (Hauskeller, 2017).

Insights articulated in the philosophy and sociology of science followed and expanded on Kuhn, such as the Mode 2 concept by Helga Nowotny and colleagues. Mode 2 conceptualizes science as inextricably interwoven with past and current social practices as well as future plans and expectations, as a complex collaborative endeavour that generates knowledge in the context of application. 'Science could no longer be demarcated from the "others" of society, culture and more arguably economy' (Nowotny et al., 2001, 1). Including these dimensions in the study of a science better captures the dynamics we want to highlight, opening up reflective and constructive studies of life science developments. What is missing is a similar inclusivity and subjection of the social sciences and humanities to said analysis of, for example, stem cell research.

The common metaphor 'field', as used widely in the history and philosophy of science, suggests that a research area can be identified through reference to its object or method. The field metaphor suggests that a distinguishable science develops in a logic of its own, driven largely by immanent factors and with definable boundaries. Yet the conduct of research is cross-sectional in and between the life sciences, humanities, and social sciences, not confining itself to integrate into a neat order.

Stem cell science represents itself as unified by the object from which it gets its name: stem cells. *Stem cells*, however, embody an obscure and transient empirical object. Plus, stem cell research has commonalities with many other life sciences in its methods and social embeddedness. Concerning its methods, stem cell research is not a distinct 'field' but a cross section relating to various disciplines with shared routines and knowledge.

Conflicting and mutually re-enforcing engagements between many pre-existing or accompanying developments in social and scientific practices and institutions are often not reflected in the narratives scientists and journals use to explain science. We problematize the marginalization of these influences in order to portray stem cell research as shaped in and by the socio-political and ethical conflicts in society.

The matrix does not require the notion of a field or similar metaphors. It emphasizes that disciplines, institutions, practices, and interests that contribute to the life sciences are shared across life science specialisms. For example, particular applications as well as moral norms and laws apply to genetics, reproductive medicine and stem cell research, which among one another exchange materials and build on one another's laboratory techniques and objects. The matrix is a discontinuous, multi-institutional, and multidisciplinary space of knowledges, interests, norms, and practices. Social, economic, and other conditions are not weighted as secondary to developments in the laboratory and clinic. They are inextricably part of the stem cell science and its objectives, as elements of its matrix. Ethics and philosophy, historical narratives and journalistic accounts, too, are constitutive and representative parts of stem cell science – they are not external to it.

A comprehensive whole of stem cell research cannot be captured because of these open boundaries and the many connections to other practices. The matrix approach opens up and includes interconnecting partial perspectives that can be held to account for their viewpoint and methods. They contribute situated knowledges (Haraway, 1988) to the stem cell discourse, for empirically informed and conceptually reflective ethics and philosophy of science.

The advantage of the matrix approach is that it can conceptualize the epistemic convergence and divergence that characterize recent life sciences (Andersen, 2016). It can recognize and attend to the tensions between plural norms and values in the moral, epistemological, and economic sense (cf. Rajan, 2006; Gottweis et al., 2009; Beltrame and Hauskeller, 2018). In the matrix we find multiple and changing loci of power and gravity. It is in flux, a multi-intersectional web of ways of knowing, doing, laboratory and governance technologies, and modes of evaluation.

The understanding of a problem affects and is affected by normative debates about different kinds of values enmeshed in it. If the object, in this case the stem cell, is itself overdetermined with conflicting value judgements and contentious in society, then commentary and judgement cannot style themselves as from the outside or the margins. Ethicists, too, often take positions implicitly, when and through choosing what they accept to be facts and decide to emphasize, thus privileging one perspective over others. Because the social science, humanities, and media debates also influence the perception and understanding of stem cell research in wider society and even to some extent in science itself, it matters that they are critically aware and reflective in their interactions and judgements. In the matrix, normativity and ethics enter at different points and in diverse ways that normative ethics especially must reflect, in view of its own role. An examination of the multiple factors shaping the development of stem cell research also has a heuristic function for ethics. Many ethical issues in the broadest sense fall into the chasm between different disciplines and tend to be overlooked by common bioethical approaches. Moral norms and values, often implicit in acts that seem not to discuss normativities beyond those directly at stake, reach further and actually build on a whole set of societal premises and conditions as well as material conditions for scientific practice. Complex implicit and explicit values that create conditions for laboratory research or legal discourse, for example, are mobilized and weighted. If certain forms of stem cell research cannot actually come into the clinic via the approved channels of scientific medicine because regulatory hurdles and conditions have aligned to form major obstacles, then research pathways dry up – whilst the unproven use flourishes in private hospitals (see Chapter 2, in this volume; Hauskeller, 2018).

Building on critiques of bioethics (Haimes, 2002; Hedgecoe, 2004; De Vries et al., 2007), the matrix approach provides a reflexive framework for responsible ethical and normative examinations of the life sciences. It can also encompass analyses of their roles in societal power and knowledge dynamics. To advance ethical reflection that can keep pace with the complexity of the tasks, the ethics within and relating to stem cell and adjacent research needs to be considered also in the context of the socio-economic interests, political influences, and value orders, and in the epistemological configurations organizing it.

In what follows we first discuss different ways of representing stem cell science in different disciplines and discourses and examine the implications of these representations for the ongoing critical discussions and ethical reflections. This section develops the matrix perspective by drawing out how it differs from existing approaches of making sense of stem cell research. In step two, we explain briefly how this perspective is illustrated across the book chapters – most of which report findings from analyses of specific practices in stem cell research, some from within, most from without the laboratory.

The chapters in this volume present different aspects within the matrix – some accepting or challenging directly the conventional definition of stem cell research as a field, others studying how the practices employed and produced in this research are adopted, reflected, and used by different professional or academic groups such as lawyers, theologians, patients, or economists. In the matrix of diverse, intersecting elements, biology and medicine are prominent, but they do not feature as the mutually interdependent but epistemically and socially independent producers of the gestalt and trajectory of the research and its uses. What we hope to show is that the conditions in and responses from different parts of society to that research have become constituent parts of its specific configurations and practices in different research projects, different laboratories and clinics, and in different countries. We believe that this characterizes the life sciences more generally, but argue for this perspective especially in relation to stem cell research.

Different approaches to conceptualizing the life sciences

Below we discuss why the matrix approach rejects forms of understanding and the presentation of stem cell research in idealized, simplistic ways, especially in what we call *origin narratives* – and how the approach builds on and integrates insights from historical accounts of how stem cell research came about, as well as philosophical and STS approaches.

'Breakthrough' myths

A conventional narrative about stem cell research, widely found in articles, textbooks, and online (Vogel, 1999), goes something like this: stem cell research emerged as a distinct domain of bio- and medical science triggered by a scientific 'breakthrough'. Usually this past event is marked with reference to a pair of articles reporting the cultivation of human stem cell lines in the laboratory (Shamblott et al., 1998; Thomson et al., 1998). The new kind of cell lines, a contemporary research object in the laboratory derived from early embryos *in vitro* using refined cell culture techniques, attracted researchers in different existing specialisms, who began to work on such human cell lines. Diverse research agendas and specialisms regrouped and rearticulated their aims and methods increasingly using the new denominator stem cell research. Twenty years after the first cultivation of hESC lines, stem cell research is a fast-paced field with a large economic turnover and, in some countries, big private markets for stem cell treatments that do not conform to established standards of safety and efficacy (Petersen et al., 2017). Of the many cell therapies envisioned, only a few have been licensed, although numerous are at different stages of clinical trial (Trounson and DeWitt, 2016). A range of new techniques and biological insights into the properties of cells and tissues *in vitro* has been created, informing regenerative medicine but also reproductive and organ transplantation medicine (Hauskeller and Weber, 2011; Zenke et al., 2018).

Such a narrative selects, orders, and interprets one or several origin events to create a coherent birth story for a new endeavour in the sciences. It suggests progress, major successes owed to identifiable specific discoveries, and reaffirms the importance of that new 'field'. It also firmly places the evolution of this 'field' as an inherent part of the dynamics within science – as a product of science.

Scientific practices and foci are in flux. Elements and techniques overlap and researchers transition from one into another (Powell et al., 2007). Objects and technologies as well as physical infrastructure and laboratory space are often shared. One research project can be framed as a contribution to several fields; for example, establishing markers for hematopoietic properties can be achieved through genomics, stem cell, or cancer research. These metaphoric 'fields' are not separated by hedges nor is ownership registered. Between the zones of overlap, crossover, and similar interests, there is also competition. A team may apply for research funding from different dedicated streams and results may be published in a range of specialist journals with different aspects of the method and findings highlighted. Specialisms emerge and regroup, coalescing around new concepts or practices. This fluidity allows the regular celebration of advances and 'breakthroughs', which in turn feeds the expectation that the life sciences are moving forward. Reasons to produce such rationales – apart from self-celebration and reaffirmation of status – are pragmatic, especially in light of the pressures to defend contested research in public and attract the required large amounts of public and private research sponsorship.

The breakthrough narrative is ideological more than explanatory. It creates the myth of both the extraordinary scientist and dramatic immanent change. The long-term labour across many laboratories and teams upon which 'breakthroughs' rest, as well as the societal contexts that shape science in practice, are not taken into account. Kuhn (1962) and Foucault (1970) have, in different ways, analysed broader conceptual shifts in how science operates, such as paradigm changes, or the order of things it studies. Ludwik Fleck has illustrated how power and community relations among teams affect what counts as proper science (Fleck, 1935). In each of these classic works science is a social enterprise, in which individuals who apply unconventional perspectives with inconvenient implications, if accepted, tend to be ignored or marginalized rather than awarded laurels, at least for a while. The narration of origin events creating stem cell research in particular has been criticized in the history, and social studies, of science (Geesink et al., 2008). The matrix approach builds on these critiques – moreover though, it problematizes the idea that science is driven, if not solely then mostly, by science per se (cf. Collins and Pinch, 1998). We develop these points in the next two sections.

Expanding multi-strand history narratives

History of science studies often situate the different disciplinary and knowledge elements of a contemporary 'field' and explore their genealogy. For example, an overview on the notion of stem cell in a science journal concludes,

> [in] summary, the early uses of the term stem cell were made in the late 19th century in the context of fundamental questions in embryology: the continuity of the germ-plasm and the origin of the blood system. The demonstration of the existence of hematopoietic stem cells ... established these cells as the prototypical stem cells: cells capable of proliferating nearly indefinitely (self-renewal) and of giving rise to specialized cells (differentiation).
>
> (Ramalho-Santos and Willenbring, 2007, 37)

It highlights the point that stem cell research absorbs different strands of nineteenth- and twentieth-century biological and medical research, depending on whether the word, the concept, or the laboratory identification of such cells is foregrounded. There is not one grand discovery. The single breakthrough from 1998 appears as just one of many elements in a complex research landscape.

Taking into account histories not of a particular disease or cell type but multidisciplinary genealogies contributing to stem cell research, the historical analyses show that no single definitive origin or founding event can be identified. Whereas the name and idea of a stem cell had been around since the late nineteenth century, the laboratory creation of the first embryonic stem cell lines, from mice, happened in 1981 and has been credited to two British teams (Evans and Kaufman, 1981; Martin, 1981). In 2007, after hESC lines had been created, the prominent Nobel Committee recognized the importance of this earlier work with an award for Martin Evans (Nobel Committee, 2007). Although it undermines birth event myths, the common type of account that lists a chain of individual discoveries and discoverers usually pays insufficient attention to extra-scientific events and conditions that drive and shape science.

Broader analyses have been provided by historians of science and medicine on individual elements such as the histories of embryology, genetics and heredity, or cancer research. Research into heredity, conceptually narrowed down to genetics, now wider again in the concept of genomics in tandem with cell biology and embryology, are all very prominent in stem cell research and highlight the latter's close entwinement with the former. Further historical studies of basic and applied research draw out a web of contributing knowledge practices that cover both epistemological and political quests, for example in Helga Satzinger's analysis of the role of political ideologies, racism, and sexism in the status and promotion of different concepts informing genetics and hormone research (2014). Concerning stem cell research directly, Beatrix Rubin (2008) and Holger Maehle (2011), among others, examined it in relation to radiation and cancer research. Understanding the response of the human organism to radiation and its often deadly effects became politically imperative from the first half of the twentieth century onwards. Scientific experimentation, medical use, and especially the use of atomic bombs created an imperative to find ways of helping those affected or at risk. From the 1950s, medical research with cultures of stem cells from bone marrow and tumours began to find ways of offsetting radiation damage, incubating the development of radiotherapy combined with the reconstitution of their blood and immune systems for some cancers (Rubin, 2008; Maehle, 2011).

The matrix approach differs from history of science narratives in its emphasis on the mutual interdependence and influence of diverse and contradictory societal and scientific developments in addition to multi-strand and non-linear scientific and political events changing sciencescapes. Stem cell research as a 'field' was politically and socially promoted, albeit as a fusion of diverse disciplinary techniques and concepts around an object within the sciences. Its contemporary gestalt has been influenced by decisions and events since the late 2000s. Those include choices and directed collaborations across political, social, and commercial sectors in the hope that this focus and coalescence of biomedical sciences would deliver societal benefits, a viewpoint expressed for instance in UK policy papers in the early 2000s (Hauskeller, 2004). The 1998 'breakthrough' may represent both the technical fine-tuning of methods of cultivating and altering embryonic cells, as well as a convenient rhetorical vantage point to make a new 'field'. Terminologically, however, it seems that instead of disciplines (biology, medicine), institutions (clinic, laboratory), or techniques (radiation, transfusion and cell transplantation, developmental microbiology), the object was chosen as identifier: the (human embryonic) stem cell. The myth of the field formation around an object is established in the naming routines in medicine, e.g. cardiology, embryology, or haematology, and has been adopted in the life sciences. Also in stem cell research, an object-based denominator has gained recognition and traction in academic and public discourses, so that historians, philosophers, and ethicists rarely critique the suggested unity of stem cell science.

Why 'stem cells' are not field-forming objects

What stem cells are is a controversial subject among philosophers and stem cell scientists alike. The special characteristics of a stem cell as commonly defined by scientists are a high proliferation rate and the capacity to divide into both cells such as itself and more differentiated daughter cells. But whether such are the properties of special cells that are naturally and continually present, often dormant, in the organism – or whether they are states of cells that many cells can potentially enter into and leave again – is matter of a long-standing controversy. Put differently, stem cell capacities may be intrinsic features of identifiable cells or rather a transient feature of many cells, a way of being adopted in response to conditions in the cell environment. How stem cells are conceptualized has implications for the feasibility of advancing therapeutic strategies, as has recently been discussed by Lucy Laplane (2016) with the example of cancer stem cells. She demonstrates a spectrum of possible metaphysical background assumptions and the relevance of clarifying these assumptions for therapeutic application. For now, consensus and clarity about stem cell concepts are lacking not only on the ontological level. The empirical concepts implicitly or explicitly guiding practitioners' identification and characterization of stem cells are manifold and not unified in an overarching concept.

The huge variety of different stem cell types, in different species, developmental stages, tissues and cell cultures, raises the issue of clear and unambiguous criteria for distinguishing stem cells from other cell types. So far, at least, specific features measurable in all but only stem cells have not been proved experimentally. Three attempts at identifying a common molecular signature of stem cells in the early 2000s failed (cf. Laplane, 2016, 115–120). This lack of a reliable characteristic of stem cells could have consequences for the unity of the 'field' of research: '[w]ithout a general "signature" of traits shared by stem cells across different environments, the field of stem cell biology seems fragmented, unified only by a common label for disparate objects of study' (Fagan, 2013, 7).

Melinda Fagan argues that shared among stem cell researchers are the basic experimental steps of extracting cells from an organismal source and determining their properties, but that knowledge about stem cells is always bound to specific experimental designs and refers not to any 'natural stem cell' but to artificial entities in specific laboratory contexts (Fagan, 2013, 68). This 'experimental relativity' (see Chapter 8, in this volume, p. 112) implies that extrapolation of experimental results is problematic and that the influence of the cell culture conditions on the measured features of cells is indeterminable: 'the investigation might be forcing the stem-cell phenotype on the population being studied' (Zipori, 2004, 876). Other epistemic problems arise from the stem cell concept itself and the basic structure of stem cell experiments such as the evidential gap between experimental data and hypotheses about stem cell capacities (see Chapter 8, in this volume).

Far from providing an empirical and unambiguous basis for the formation of a new research field, thus far 'stem cell' is a messy concept referring to not-well-defined objects with fuzzy boundaries. That which is available is only a patchwork of highly context-specific knowledge about different groups of cells that are mostly research and laboratory artefacts. The matrix approach is distinct from many traditional forms of social studies of science that 'tended toward the dissolution of all distinctive boundaries demarcating the sciences' (Keller, 1992, 3). In accordance with Evelyn Keller's 'middle-of-the-road-position' (ibid.), regarding stem cell research as a social practice does not mean that inner-scientific dynamics and epistemic commitments do not merit close scrutiny. The models and explanatory standards accepted within a research community, as well as technological and further conditions that shape the course of science are part of matrix studies. History and philosophy of science are traditionally bound to historical biographies of scientific objects (Arabatzis, 2006) and to models and techniques (Chang, 2004) or research groups (Kohler, 1994). Rita Temmerman has demonstrated how terminological conventions bury multi-layered intersections of conceptualization, also in the life sciences (Temmerman, 2000, 103). Moreover, Ann-Sophie Barwich has argued that once philosophy and history of science look at the latter as a complex of changing material practices, instruments, discourses, etc., an extension is needed of the meaning of terminology that refers to concepts, models, objects, and discourses in the sciences (Barwich, 2013). The matrix approach addresses these shortcomings, integrating technical and conceptual details as analysed by Fagan and Laplane, with knowledge of the web of social, political, economic, and ideological factors and institutions.

Why methods and clinical goals do not distinguish a stem cell 'field'

Notwithstanding the outlined epistemological problems, the stem cell lends a name to a diverse and sprawling multitude of research endeavours. The spread of scientific pathways taken, and the researchers involved, is evidenced by the dedicated science journals and general medical and biology journals in which relevant publications appear.

The established organization of science into distinct disciplines and specializations, each with its specific viewpoint, communities, and institutional infrastructures, is modelled around objects and technologies. In this logic it seems reasonable to assume that a new kind of object and affiliated set of technologies can disrupt this order, bringing forth a new field that draws on and pulls in expertise as well as infrastructures across several fields of research. The researchers working on stem cells in evolutionary and developmental biology, in oncology, haematology, cell biology, and in reproductive medicine – providing the cells for making stem cell lines – were and still are organized in different disciplines, departments, and bound up with different institutional contexts. They did not work on a common project, but rather provided tools, material, and techniques taken up in research with stem cells or tissue engineering.

The descriptor stem cell research gained traction, and the 'field' gained reality in the sociological pragmatic sense, in the early 2000s when some biologists began to identify their professional expertise as stem cell researchers rather than staying aligned with the previous disciplinary order. Specialized training programmes in biology departments, dedicated journals, teaching books, education courses (Hauskeller, 2004), as well as academic conferences and scientific societies are other elements in the institutionalization of a stem cell research infrastructure – the International Society of Stem Cell Research (ISSCR) was founded in 2002, for instance. Common pragmatic interests favoured the organization of a unifying profile under the new umbrella term, because it opened up multiple existing governmental and charity routes to funding for research, including funds dedicated to treating specific diseases (e.g. cancer, motor neurone disease, heart disease). This also entails that an imperative to translate research from bench to bedside shapes scientists' behaviour and the accounts of their work. The directions and pathways in basic research tend to be justified with reference to new medical treatments.

Many researchers feel that a 'translational imperative' (Harrington and Hauskeller, 2014) requires them to find ways of justifying their work with its potential societal benefit. Other critics of this pressure have argued that the rebranding of basic research as working towards regenerative medicine raised unrealistic expectations that cures for a broad range of diseases would become imminently available (Kitzinger, 2008; Petersen et al., 2017). It has also been argued that the clinical goals were crucial to the rapid expansion of stem cell research and laboratories worldwide, and shaped the development of standards and methods of experimentation and explanation (Fagan, 2013, 234). Whether this is a less diffuse bond between diversely organized and oriented research teams across the world has also to be put into question. The orientation towards future clinical utility, whilst shared among stem cell researchers, is not something that necessarily fosters a sense of community among them. The translational imperative affects all basic life science research. Stem cell research competes with other approaches in genomics, nano-medicine, and so forth, for funding – and also for the attention of medical practitioners, whose enthusiasm is needed for the protracted route into the clinic via clinical trials or other experimental uses. The possible bio-economic utility and patentability of research findings (see Chapters 3, 4, and 12, in this volume) may increase competition rather than community. Thus it is as easy to argue that it dis-incentivizes (self-)critical assessments of the research conducted, and that it has led to a deficit in critique, honesty, and transparency. The pressures of delivering timely results have contributed to premature publication of laboratory results and some of the many instances of scientific fraud in stem cell research (e.g. Kim, 2013). In this volume, Antonakaki presents a recent case study on the technical and political aspects of the persistent problems with establishing methods to counteract fraud and to verify experimental results.

The matrix of disciplines, institutions, interest groups, and individual agents that take an interest in stem cell research can be studied as a configuration of multiple points of interaction and tension. Different standpoints and research trajectories can be taken to examine what stem cell science is in its different configurations of interacting discourses. The many empirical axes of study problematize diverse nodes of contention concerning concepts, normative tensions, economic aspects, or diverse research pathways. We propose adopting an epistemological perspective that sees responsibility for the forms of stem cell research that happen and what happens in and with them, lying not only with the natural scientific and medical contributions to stem cell research. Ethicists, lawyers and policy-makers, economic and industrial agents as well as patients and consumers carry responsibility for their contributions to what and how stem cell science happens. In the following sections we introduce the chapters in this volume in relation to the matrix approach.

Contributions and organization of this volume

This book presents original perspectives on the current state of stem cell science from a range of disciplines, including philosophy and various social sciences as well as laboratory research. The chapters do not themselves take the matrix perspective or question how stem cell research has been defined. They include state-of-the-art accounts that detail specific themes such as tensions over epistemology and scientific practices. Other chapters analyse cases or uses of stem cell research that illustrate the peculiar discrepancies in political aims or moral commitments that shape its societal and scientific practices. The practices studied in the stem cell matrix concern private and public enterprises, partnerships between them, patient organizations, regulatory authorities, scientific concepts, as well as political and academic interests. The disciplinary contexts in which the participants work and their foci of attention and expertise jointly show a matrix of points of interest and contentions in flux. Several chapters reflect on the promises of biomedicine to deliver health benefits that attract both investor and patient interest in specific situations. Complex bioeconomies have grown around stem cell research, driven by the institutions that manage private or public biobanks or industries searching for new medical products. The well-studied publicity and transparency of these conditions in and around stem cell science make it a good case study to show the relevance of the matrix approach for conceptualizing the life sciences.

Together the chapters bring out the diversity of the understandings of stem cell research in different partaking disciplines and how academic discourses frame issues. A self-recursive figure becomes apparent through this multitude of perspectives: the current stem cell 'field' notion depends on putting the cells and those who work with them at the centre, and present philosophical epistemology, sociological studies of scientific or societal practice, and legal and ethical debates, as gravitating around this object. The notion of disruptive technologies has taken hold especially in debates about social media, but many practices in biomedicine and routine parts of stem cell research challenge laws, societal conventions, and moral understandings. Patient rights and clinical research regulations stand alongside laws and norms about the use of human cells and embryos for research. The configuration of viewpoints in a flexible matrix highlights various interacting interpretations of what matters about stem cell science. The cells and the clinical orientation are only aspects of broader societal dynamics that shape and respond to what happens in the biological and medical sciences.

The first three chapters investigate societal practices directly related to and affecting the research, thereby illustrating how conditions and actions from outside the laboratory shape stem cell research. They offer insights from studies of medical markets and biobanking industries as well as patenting law. *Claire Tanner, Alan Petersen, Casimir MacGregor*, and *Megan Munsie* present findings from a sociological case study of the X-Cell Center in Germany. This case was among their research cited to explore the manifold conditions enabling stem cell tourism. They highlight the roles of regulatory loopholes, the publicly nurtured problematic belief in the healing power of stem cells, low risk perception, and the appealing treatment conditions on offer. In the next chapter *Lorenzo Beltrame* analyses how economic value is produced in new forms of biobanking through the commodification of biological materials, the assetization of knowledge and technological capacities, and the exploitation of donors. He compares these aspects across two case studies – cord blood stem cell banking, and the circulation of hESC lines. *Fruszina Molnár-Gábor* examines limitations on patentability of results from hESC research through opening clauses of international and national patent laws. Her analysis of how these opening clauses, such as 'accepted principles of morality', are interpreted through case law at EU and member state levels reveals how these legal texts draw in and refer to extra-legal motivations for limits on patentability. The relationship between legal norms and extra-legal ethical standards is problematized in the end, especially concerning questions of the legitimacy of deciding upon morally contested issues by incorporating them into positive law.

The next three chapters present the social practices in laboratory stem cell research and how its internal logic works in relation to its objects, success indicators, and external conditions and pressures. Stem cell laboratory scientists present accounts of their contemporary research and its objects, a discussion that will be complemented by two philosophy of science studies on those issues. Stem cell research is thriving and has diversified its techniques and ways of achieving clinical benefits. In order to illustrate this expansion and diversification of the stem cell sciencescape, the three chapters by laboratory scientists all belong in one sense to the same specific subfield: the study of neuronal cells and neurodegenerative disorders. Yet they present different approaches including induced pluripotent stem cells (iPSC) in clinical translation, new methods of reprogramming cells *in vivo*, that is without removing them from the body, and research on 'diseases in a dish'. First *Stephanie Sontag* and *Martin Zenke* report on the current state and pitfalls of research with iPSC and the prospects for clinical application; they also present a brief introduction to the biology and experimental history of stem cell research from their point of view. *Maryam Ghasemi-Kasman* introduces the methods, advantages, and challenges of her current research on developing a way of reprogramming *in vivo* glial cells into neurons – changing cells without removing them from the organism with the aim of repairing the central nervous system from within. And finally *Irina Prots, Beate Winner* and *Jürgen Winkler* introduce their ongoing research of stem cell-based models of human neurodegenerative diseases with the aim of better understanding disease mechanisms and developing new therapeutics. They create stem cell cultures out of cells taken from patients with the respective diseases on the basic assumption that the molecular properties of the cell culture in the laboratory correspond to those in the cells of an organism with manifest disease symptoms.

The reports by laboratory researchers on their objects, objectives, and the state-of-the-art understanding of basic phenomena and clinical aspirations are complemented by the two following chapters by philosophers of science who aim to clarify the core concept in stem cell research. With her minimal stem cell model and its relation to experimental practices, *Melinda Fagan* provides an overview of the background assumptions about biological development and discusses epistemological problems, especially the evidential constraints of identifying stem cells, in light of recent debates within the philosophy of biology. Building on Fagan's insights, *Anja Pichl* then gives an explanation of how the common understanding of stem cells as clearly identifiable entities with certain intrinsic properties can still be influential on both science and society despite being epistemically untenable. The problematic societal and scientific effects of reducing complexity to retain the unifying narrative are discussed and traced back to two constituents of the field: clinical goals and methodological reductionism.

The subsequent three chapters revisit stem cell research as social practice from social science perspectives highlighting different normativities.

The chapters present perspectives from STS, philosophical ethics, and social anthropology, each illustrating the potential of reflective cooperation among the humanities, empirical social sciences, and STS for examining the intersection of knowledge practices and norms of truth and ethics. *Melpomeni Antonakaki* has studied how the scientific community intended to establish the truth and attributed personal responsibility for suspected misconduct concerning the so-called STAP (stimulus-triggered acquisition of pluripotency) technology to create pluripotency in cells. Findings from her laboratory ethnography present a detailed account of the creation of scientific facts and the establishing of credibility through record keeping and audit practices. The roles that social status and hierarchies play, as well as the observation and transparency of laboratory practices in reinforcing credibility and trust that are discussed, undercut the myths of science developing in ways driven all from within itself and the objectivity of data. Investigations of global governance discourses emphasize the fragility of the epistemic configuration and the multiple societal domains life science research is entangled with in different contexts of societal values and power orders. *Ahmet Karakaya* decentres the stagnant debate in many, predominantly

Western, countries about the moral status of the human embryo in relation to its use in hESC research. He conducted, and reports from, an empirical interview-based study on the ethical positions of Muslim scholars in Turkey on research with human embryos and the normative principles to which they refer. The predominant position among these interviewees is more liberal and expresses greater regard for the well-being of social persons than the contravening powerful arguments put forth by, for example, Catholic theologians on this issue (Levada and Ladaria, 2008). This highlights the implicit biases created in uncritical hegemonial globalized ethics about the normative foci in the moral reasoning of diverse cultures, under conditions of moral pluralism. *Achim Rosemann* shows that particular forms of governance can lead to injustices such as exclusion from new forms of business and innovation opportunities for researchers and technology producers from low- and middle-income countries. Globally distributed rules and regulations, usually designed by predominantly Western industrial and scientific institutions, have produced technical and ethical standards addressing Western moral themes such as the moral status of the embryo (see Chapters 3 and 11, in this volume; Bender et al., 2005). Hauskeller argued that these regulations tend to concur with market prerogatives of major laboratory equipment producers and the medicines industry that focus biomedical research on alleviating suffering from the diseases of the wealthy. Formal policies and regulations shape which research is conducted and how, often in unexpected ways that contradict the stated moral and legal justifications for said regulation and the stated rationales for the public investments that enable the research. Some potential avenues of research, such as therapies using autologous bone marrow stem cell grafts, are positively disadvantaged in this regulatory set-up, contrary to ethical commitments expressed by those regulators (Hauskeller, 2018). The current governance frameworks require that those who want to participate in this research must work in high-tech laboratories and evidence high standards of scientific as well as administrative expertise. Those assets and conditions are unevenly distributed across the globe. *Rosemann* examines the tensions between the management of technology risks and the demands for social justice in emerging international governance frameworks for stem cell and gene drive research. The social and ethical problems of containing genetic alterations and the risks they pose are just one example for the technological closeness of biomedical research domains and the challenges of regulating genomic technologies in and beyond stem cell laboratories.

The chapters contribute original work to their authors' respective primary disciplines, and in combination point out areas where conflicts arise between their discourses. They demarcate intersections that are part of the current stem cell matrix, without representing a comprehensive picture – acting as spotlights on diverse and contradictory societal and epistemic factors, dynamics, interests, and values that are active in the development of stem cell research. The variety of viewpoints and norms they emphasize sheds some light on ethical issues that are easily overlooked because they fall between disciplinary chairs, metaphorically speaking. Some also exemplify the case that ethics as an interdisciplinary enquiry of explicit and implicit normative assumptions of factual, sociological, philosophical, and legal studies is underway at disciplinary cross sections and that philosophical and STS can contribute critical normative work by concentrating on these.

Questioning what the focus of analysis is in order to define one's referent – or, as we have seen above, challenge the existence of a clearly defined field of stem cell research – is important for situated, self-critical reflection in philosophy and ethics. If we assume that science develops of its own accord, philosophy and ethics are commentating outsiders. We have argued that this is a false representation of the genealogy and situation of stem cell research. The sciences, natural as well as social, have a continually reaffirmed function in a society in which knowledge informs complex socio-political negotiations and institutions and vice versa. Based on this understanding, philosophers and ethicists are part of the producers and of the production process that carves out the niches and functions of science in society. Because of the publicly conducted controversies about it in the 2000s, stem cell research provides a very effective case study to unpick depoliticized linear narratives of scientific development that cover up the complex processes that go into the creation of the sciencescape. The matrix of stem cell science contains the discourses of ethicists, lawyers, philosophers, clinical and laboratory scientists, social scientists, and theologians – especially where they affect policy and the public understanding of what goes on in scientific research. It also considers the influences from non-academic participants, such as publics and the industries that contribute to the making of stem cell science and its applications.

This volume elucidates some of the constitution conditions of stem cell research to open up dialogues with scholars from all disciplines interested in furthering the understanding of the contemporary life sciences. On that basis a critical understanding of stem cell research as a social practice can emerge, which highlights the historic, societal, economic, cultural, ideological, epistemic, and material conditions, and thus the changeability of the factors shaping science pathways. Drawing on and including concepts from STS the matrix approach points philosophy of science and ethics towards ways of rethinking the science and society intersection and reshaping it conceptually. Unpacking some of the black boxes concerning constituents of the field of stem cell research raises questions about what societal goals diverse social practices including the life sciences aim to achieve. In that sense, we collated this volume to present the state of the art in reflecting on stem cell research that might contribute to collectively reorienting such work towards context-sensitive ethical and political reflections.

Acknowledgements

This matrix approach, as conveyed through the format of this book, evolved from the editors' continuing discussions following the international summer school 'Pluripotent Stem Cells: Scientific Practice, ethical, legal, and social commentary' in late 2015. Several of the contributors to this volume presented at the event, whereas others were invited to contribute later, and we are grateful to all the authors for their willingness to provide original articles on their special research projects in the stem cell matrix.

We also express our gratitude to the institutions who supported us through this process, among them first and foremost the German Federal Ministry of Education and Research (BMBF) who awarded Arne Manzeschke generous funding for organizing the summer school in Bavaria and the project's public closing event in April 2017 as well as for the edition of this volume in English (funding code FKZ: 01GP1482). The Bavarian research network on human induced pluripotent stem cells, ForIPS, funded by the Bavarian State Ministry of Science and the Arts (funding code D2-F2412.26), in which Arne and Anja were active between 2014 and 2017, also supported the project, and so did the University of Exeter, with top-up funding for copy-editing. We further thank the conscientious peer reviewers who supported the development of the chapters and, last but not least, we endorse and thank Dr Peter Hughes for his thoughtful editing, close attention to detail, and timely delivery of copy-edited texts – whilst concurrently completing his philosophy PhD.

References

Andersen, H. (2016). Collaboration, interdisciplinarity, and the epistemology of contemporary science. *Studies in History and Philosophy of Science*, 56, pp. 1–10.

Arabatzis, T. (2006). *Representing Electrons. A Biographical Approach to Theoretical Entities*. Chicago: University of Chicago Press.

Barwich, A. S. (2013). A pluralist approach to extension: the role of materiality in scientific practice for the reference of natural kind terms. *Biological Theory*, 7, pp. 100–108.

Beltrame, L. and Hauskeller, C. (2018). Assets, commodities and biosocialities. Multiple biovalues in hybrid biobanking practices, *TECNOSCIENZA. The Italian Journal of Science & Technology Studies*, 9(2), pp. 5–31.

Bender, W., Hauskeller, C. and Manzei, A., eds. (2005). *Crossing Borders. Ethical, Legal, Economic and Religious Views on Stem Cell Research. Global Perspectives*. Münster: Agenda Verlag.

Chang, H. (2004). *Inventing Temperature. Measurement and Scientific Progress*. Oxford: Oxford University Press.

Collins, H. and Pinch, T. (1998). *The Golem: What Everyone Should Know About Science*. 2nd ed. Cambridge: Cambridge University Press.

De Vries, R., Turner, L., Orfali, K. and Bosk, C. (2007). Social science and bioethics: The way forward. In: R. DeVries, L. Turner, K. Orfali and C. Bosk, eds., *The View from Here: Bioethics and the Social Sciences*. Oxford: Wiley Blackwell, pp. 1–12.

European Commission (2009). EUR 23616 EN – Global Governance of Science. Report of the Expert Group on Global Governance of Science to the Science, Economy and Society Directorate, Directorate-General for Research, European Commission. Luxembourg: Office for Official Publications of the European Communities.

Evans, M. J. and Kaufman, M. H. (1981). Establishment in culture of pluripotential cells from mouse embryos. *Nature*, 292, pp. 154–156.

Fagan, M. B. (2013). *Philosophy of Stem Cell Biology. Knowledge in Flesh and Blood.* London: Palgrave Macmillan.

Fleck, L. (1935/1979). *Genesis and Development of a Scientific Fact.* In: T. J. Trenn and R. K. Merton, eds., Chicago: University of Chicago Press.

Foucault, M. (1970). *The Order of Things.* London: Pantheon Books.

Foucault, M. (1997). Polemics, politics, and problematizations. In: P. Rabinow, ed., *Ethics: Subjectivity and the Truth. The Essential Work of Foucault 1954–1984, Vol. I.* New York: The New Press, pp. 111–119.

Geesink, I., Prainsack, B. and Franklin, S. (2008). Stem cell stories 1998–2008. Guest editorial. *Science as Culture*, 17(1), pp. 1–11.

Gottweis, H., Salter, B. and Waldby, C. (2009). *The Global Politics of Human Embryonic Stem Cell Science. Regenerative Medicine in Transition.* London: Palgrave Macmillan.

Haimes, E. (2002). What can the social sciences contribute to the study of ethics? Theoretical empirical and substantive considerations. *Bioethics*, 16(2), pp. 89–113.

Haraway, D. (1988). Situated knowledges: The science question in feminism and the privilege of partial perspective. *Feminist Studies*, 14(3), pp. 575–599.

Harrington, J. and Hauskeller, C. (2014). Translational research: An imperative shaping the spaces in biomedicine, *TECNOSCIENZA. The Italian Journal of Science & Technology Studies*, 5(1), pp. 191–201.

Hauskeller, C. (2004). How traditions of ethical reasoning and institutional processes shape stem cell research in the UK. *Journal of Medicine and Philosophy*, 29(5), pp. 509–532.

Hauskeller, C. (2017). Can harmonized regulation overcome intra-European differences? Insights from a European Phase III stem cell trial. *Regenerative Medicine*, 12(6), pp. 599–609, online first on 4 Oct., doi: 10.2217/rme-2017-0064.

Hauskeller, C. (2018). Between the local and the global: Evaluating European regulations of stem cell regenerative medicine. *Perspectives in Biology and Medicine*, 61(1), pp. 42–58.

Hauskeller, C. and Weber, S. (2011). Framing pluripotency. IPS cells and the shaping of stem cell science. *New Genetics and Society*, 30(4), pp. 415–431.

Hedgecoe, A. (2004). Critical bioethics. Beyond the social science critique of applied ethics. *Bioethics*, 18(2), pp. 120–143.

Keller, E. F. (1992). *Secrets of Life Secrets of Death: Essays on Language, Gender, and Science.* New York: Routledge.

Kim, L. (2013). Denotation and connotation in public representation: Semantic network analysis of Hwang supporters' internet dialogues. *Public Understanding of Science*, 22(3), pp. 335–350.

Kitzinger, J. (2008). Questioning hype, rescuing hope? The Hwang stem cell scandal and the reassertion of hopeful horizons. *Science as Culture*, 17(4), pp. 417–434.

Kohler, R. E. (1994). *Lords of the Fly. Drosophila Genetics and the Experimental Life.* Chicago: University of Chicago Press.

Kuhn, T. S. (1962/2012). *The Structure of Scientific Revolutions.* 50th Anniversary Edition with an Introductory Essay by Ian Hacking. Chicago: University of Chicago Press.

Laplane, L. (2016). *Cancer Stem Cells. Philosophy and Therapies.* Cambridge: Harvard University Press.

Levada W. and Ladaria L. F. (2008). Congregation for the Doctrine of the Faith. Instruction *Dignitas Personae* on Certain Bioethical Questions, [online]. Available at: www.vatican.va/roman_curia/congregations/cfaith/documents/rc_con_cfaith_doc_20081208_dignitas-personae_en.html [Accessed 25 Jan. 2019].

Maehle, A.-H. (2011). Ambiguous cells: The emergence of the stem cell concept in the nineteenth and twentieth centuries. *The Royal Society Journal of the History of Science*, 65(4), pp. 359–378.

Martin, G. R. (1981). Isolation of a pluripotent cell line from early mouse embryos cultured in medium conditioned by teratocarcinoma stem cells. *Proceedings of the National Academy of Sciences USA*, 78, pp. 7634–7638.

Nobel Committee (2007). Advanced information. NobelPrize.org. Nobel Media AB 2019, [online]. Available at: www.nobelprize.org/prizes/medicine/2007/advanced-information/ [Accessed 9 June 2019].

Nowotny, H., Scott, P. and Gibbons, M. (2001). *Re-Thinking Science. Knowledge and the Public in an Age of Uncertainty*. Cambridge: Polity Press.

Petersen, A., Munsie, M., Tanner, C., MacGregor, C. and Brophy, J. (2017). *Stem Cell Tourism and the Political Economy of Hope*. London: Palgrave Macmillan.

Powell, A., O'Malley, M. A., Müller-Wille, S., Calvert, J. and Dupré, J. (2007). Disciplinary baptisms: A comparison of the naming stories of genetics, molecular biology, genomics, and systems biology. *History and Philosophy of the Life Sciences*, 29(1), pp. 5–32.

Rabinow, P. (2003). *Anthropos Today. Reflections on Modern Equipment*. Princeton/ Oxford: Princeton University Press.

Rajan, K. S. (2006). *Biocapital. The Constitution of Postgenomic Life*. Durham: Duke University Press.

Ramalho-Santos, M. and Willenbring, H. (2007). On the origin of the term "Stem Cell". *Cell Stem Cell*, 1(1), pp. 35–38.

Rubin, B. P. (2008). Therapeutic promise in the discourse of human embryonic stem cell research. *Science as Culture*, 17(1), pp. 13–27.

Satzinger, H. (2014). The politics of gendered concepts in genetics and hormone research in Germany, 1900–1940. *Gender History*, 24(3), pp. 735–754.

Shamblott, M. J., Axelman, J., Wang, S., Bugg, E. M., Littlefield, J. W., Donovan, P. J., Blumenthal, P. D., Huggins, G. R. and Gearhart, J. D. (1998). Derivation of pluripotent stem cells from cultured human primordial germ cells. *Proceedings of the National Academy of Sciences USA*, 95, pp. 13726–13731.

Temmerman, R. (2000). *Towards New Ways of Terminology Description: The Socio-cognitive Approach*. Amsterdam: Benjamins.

Thomson, J. A., Itskovitz-Eldor, J., Shapiro, S. S., Waknitz, M. A., Swiergiel, J. J., Marshall, V. S. and Jones, J. M. (1998). Embryonic stem cell lines derived from human blastocysts. *Science,* 282, pp. 1145–1147.

Trounson, A. and De Witt, N. D. (2016). Pluripotent stem cells progressing to the clinic. *Nature Review Molecular Cell Biology*, 17(3), pp. 194–200.

Vogel, G. (1999). Breakthrough of the year. Capturing the promise of youth. *Science*, 286(5448), pp. 2238–2239.

Zenke, M., Marx-Stölting, L. and Schickl, H., eds. (2018). *Stammzellforschung. Aktuelle wissenschaftliche und gesellschaftliche Entwicklungen*. Berlin-Brandenburgische Akademie der Wissenschaften. Baden-Baden: Nomos.

Zipori, D. (2004). Opinion: The nature of stem cells: State rather than entity. *Nature Reviews Genetics*, 5, pp. 873–878.

2 Big promise, big business

The sociocultural and regulatory dynamics of stem cell tourism

Claire Tanner, Alan Petersen, Casimir MacGregor, and Megan Munsie

Abstract

Stem cell tourism is a global phenomenon. People travel across geographical borders and jurisdictions to buy unproven stem cell 'treatments'. We present a sociological case study, the X-Cell Center in Germany – once a popular destination of choice for stem cell travellers, shut down by regulators in 2010 after several deaths occurred. We offer a perspective on the conditions that enable stem cell tourism including public perceptions of safety, national stereotypes, personalized care offered to clients, ambiguous regulation and public belief in the power of stem cells.

Introduction

The growth of investment and research in stem cell science has been accompanied by a significant growth in the market of purported stem cell treatments (SCTs). In recent years, the phenomenon of stem cell tourism[1] has emerged as an issue of scientific and regulatory concern. Stem cell tourism refers to a global trend whereby patients and carers travel across national borders and jurisdictions to purchase, sometimes at considerable cost, claimed SCTs for which there exists little to no clinically verified evidence of safety or benefit (Petersen et al., 2017). This chapter considers the conditions that enable commercial stem cell businesses to flourish through an examination of the rise and fall of the X-Cell Center in Germany. Adopting a sociological approach, we consider how the fastest growing markets are currently built and sustained. In particular, we consider how a combination of regulatory ambiguity and/or loopholes, direct-to-consumer marketing which exploits public perceptions of high national standards in regulation and medical care where clinics are located, combined with the sociocultural belief in the power of stem cells, as well as tailored one-on-one 'care' for clients, create the conditions for businesses to flourish.

Although the exact beginnings of stem cell tourism are ambiguous, one early media article reported that more than forty patients – mostly Americans – had received 'stem cell therapy' in a Bangkok hospital in 2005 (Montlake, 2006). The procedure – presented as 'something out of the ordinary' and not available 'at home' – was reported to involve stem cells that were grown from the patient's own blood and then processed in a laboratory in Israel before being flown back to Bangkok and injected into the patient's heart (Montlake, 2006). Since this report, numerous studies have identified a growing number of commercial clinics promoting SCTs via online direct-to-consumer marketing (Lau et al., 2008; Ogbogu et al., 2013; Connolly et al., 2014). The most recent studies have identified hundreds of stem cell clinics operating around the globe (Berger et al., 2016; Turner and Knoepfler, 2016).

These commercial stem cell clinics claim to use cells that are autologous (from the patients' own body) or derived from cord blood, human foetuses, or embryos. The cells are commonly injected into the body intravenously, intramuscularly, and/or by lumbar puncture, or use a combination of all these methods of delivery for people depending on their condition (Petersen et al., 2013, 2017). According to their online advertisements, some clinics have even reported to have administered cells directly into the brain or spinal cord (Lau et al., 2008). Most clinics adopt a 'one-size-fits-all' approach, promising that their treatments have the potential to benefit people with a wide range of conditions and diseases including erectile dysfunction, diabetes, heart disease, autism, cerebral palsy, multiple sclerosis, motor neuron disease, Alzheimer's disease, spinal cord injury, and blindness. There are large variations in treatment regimes for different clinics, with some clinics offering one-off treatments and others offering twelve-week courses involving live-in accommodation, daily and weekly SCTs delivered multiple ways accompanied by nutritional supplements, and intensive physiotherapy of up to five hours a day with fees ranging from 3000 to 66 000 (EUR) (Petersen et al., 2017).

The clinics also use a variety of online marketing tools to appeal to prospective patients including patient stories, blogs, YouTube videos, and links to scientific and news sources. Emotive anecdotal stories are commonly drawn upon to show the potential benefits of treatment and the prospect of significant improvement for those people with the strength and determination to undergo treatment (Petersen and Seear, 2011; Connolly et al., 2014). Whilst scientific, medical, and regulatory communities are concerned by the growing number of clinics that offer 'expensive, cell-based therapies that are biologically implausible, known to be ineffective, or have no proven benefit' (Ikonomou et al., 2016), for many patient activists there is concern that more stringent regulations will impinge on their right to access experimental treatments on compassionate grounds (Adriance, 2014; Petersen et al., 2017).

Whilst it has been possible to illustrate a rapid growth in this controversial sector, documenting the full extent of the stem cell tourism industry is extremely difficult. This is because there is presently no mechanism for, or governance framework requiring, citizens to report medical travel, nor for businesses and clinics to collect data on and or report their operations or the number of patients they are treating. However, while stem cell clinics have been located in jurisdictions with lax or absent regulatory frameworks and/or enforcement – such as India, Thailand, Russia, Mexico, Ukraine, and China – increasingly they have emerged in countries perceived as having both stringent regulatory standards and high standards of medical care, such as Germany, Australia, and the United States (Berger et al., 2016; Turner and Knoepfler, 2016). This development can be at least partly explained by jurisdictional variations in regulatory provisions and enforcement with respect to the use of autologous cells (cells taken from and re-administered to the patient following their manipulation) (Munsie and Pera, 2014; McLean et al., 2015; Turner and Knoepfler, 2016). The X-Cell clinic in Germany offers an apposite example in this respect. Exploiting a period of regulatory ambiguity in Germany around the use of autologous cells, the X-Cell clinic was, ironically, able to exploit the public image of Germany as a nation with stringent standards in both medical innovation and care, in order to market and sell SCTs to patients.

In this chapter we offer a sociological perspective on the dynamics that have shaped and sustained the development of the stem cell tourism phenomenon, and especially on the recent growth of clinics claiming to sell autologous SCTs to patients. In so doing, we map the confluence of factors that influence patient perceptions and the expansion and contraction of the market in different jurisdictions over time. These factors include the changing of regulatory environments, patient advocacy, media reporting of stem cell breakthroughs and stem cell tragedies, direct-to-consumer advertising, and wider politico-economic processes. As mentioned, we focus our attention on a single case study, the X-Cell Center in Germany – once a popular destination for stem cell travellers that was eventually shut down by regulators after a number of adverse events involving patients. In particular, we draw upon in-depth qualitative interviews with patients and carers who have considered or received SCTs in Germany – as well as media reports, direct-to-consumer advertising, and the perspective of German regulators.[2] Through our analysis we demonstrate how people's perceptions of, and hopes and investments in stem cells, and the clinics that offer them, are shaped by popular sociocultural constructions of stem cells as well as the actions of a range of constituents, including scientists, clinicians, regulators, the biotechnology industry, governments, and the media. In so doing we argue that there is a need for greater understanding of the sociocultural and regulatory dynamics underpinning the SCT marketplace that have the capacity to enable and potentially contain stem cell tourism.[3]

The rise of the X-Cell Center

The X-Cell Center (X-Cell), at the Eduardus Hospital in Cologne in Germany, first opened its doors and treating rooms to patients in 2007, later extending its operations to the Dominikus Hospital in Düsseldorf. The clinic claimed to offer treatments for patients 'suffering from diseases for which standard medical treatment has no cure as yet', including Parkinson's, Alzheimer's, cerebral palsy, multiple sclerosis, diabetes, heart disease, spinal cord injury, macular degeneration, and autism (X-Cell Center, 2007). Patients were charged between 3000 and 10 000 (EUR) per SCT. Treatments involved bone marrow extracted from the patient, the purported stem cells isolated and then administered back to the patient intrathecally (via lumbar puncture into the spinal cord), intravenously (via drip), or intramuscularly (injected into the muscles). Depending on the condition, the cells of some patients were injected directly into the brain. Whilst the use of hematopoietic stem cell transplantation, where the cells are obtained from bone marrow or cord blood, is a proven and routine treatment for some immune and blood disorders, such as acute myeloid leukaemia and lymphoma (Copelan, 2006), there is limited scientific rationale and evidence for the clinical application of bone marrow-derived stem cells for other conditions and/or diseases (Daley, 2012; Trounson and McDonald, 2015).

Despite initially being banned in the Netherlands (Sheldon, 2007), X-Cell was able to successfully operate due to ambiguity in German and EU regulations. In particular, there was ambiguity with respect to whether the services offered by X-Cell should be considered a medical practice or regulated as a therapeutic product. Notably, under European law doctors are allowed to treat patients in European hospitals, under a 'hospital exemption', when a therapy is in the process of being certified, clinically tested or used as a treatment of last resort, subject to the approval of a relevant federal authority. Furthermore, whilst it could be considered that the SCTs administered at X-Cell were an Advanced Therapy Medicinal Product (ATMP) and therefore under the remit of the European Medicine's Agency and relevant national authorities, this legislative change did not come into effect in Germany until December 2008. The thirty-six-month transition period contributed to confusion as to whether a stem cell-based product had to be regulated under the EU regulation or German Drug Law (Schnitger, 2014). There was also significant debate around the definition of a stem cell-based product which compounded existing jurisdictional and regulatory uncertainty around the practices of X-Cell (Petersen et al., 2017, 103–105).

Whilst initial regulatory ambiguity enabled the clinic to operate and was a significant factor shaping practices, such uncertainty is in itself unlikely to sustain a market of SCTs. If clinics are to attract patients, they need to advertise their products and their benefits – and this generally occurs via the Internet.

Building a market

For X-Cell, like many other SCT businesses operating around the globe, online direct-to-consumer marketing was used to attract international patients over the course of the clinic's operation from 2007 to 2011. In 2007 the front page of the Center's website read (with the facility of translation into three languages: German, English, and Dutch):

Are you suffering from a serious condition that has failed to respond to conventional treatment? Your chance: Stem cells from your own body (autologous stem cells) X-Cell Center is the first privately owned center in Europe specializing in using adult stem cells for the purposes of regenerative medicine. By using adult stem cells, which are taken from the patient's own body, prepared, and then re-injected, our specialists are aiming at treating severe diseases. Stem cell therapy is promising for patients who are suffering from diseases for which standard medical treatment has no cure as yet. We apply the highest technological and medical standards in the world.

(X-Cell Center, 2007)

By 2010 the website offered (with the facility of translation into fifteen languages) both text-based testimonials from patients with a wide variety of conditions as well as photographs of patients and patient videos testifying to the benefits of treatment, the professionalism of the clinicians, and the superior care provided by X-Cell. For example, a screenshot of the website in December 2010 reflects the powerful use of patient stories which attracted thousands of paying international customers to X-Cell, numbering more than 13000 for a two-year period at its zenith, according to a lawyer who investigated the Center's operations (Petersen et al., 2017, 106). Glowing testimonials line the left-hand side of the webpage. These include: 'VIDEO – Happy Holidays 2010 – Successfully treated stem cell therapy patients hailing from around the globe share their pictures and comments'; 'See her walk again'; 'Stem Cell Therapy in Germany Results in Significant Improvements for Teenage Cerebral Palsy Patient'; 'Secondary Progressive MS Patient Improves Dramatically'; 'US Heart Failure Patient Considers Himself "Cured"'; 'Remarkable progress of Caroline' (X-Cell Center, 2010). The use of patient testimonials continues to be a key marketing strategy adopted by clinics globally (Ogbogu et al., 2013). However, our research indicates that current marketing practices adopted by some clinics also involve more aggressive techniques, including the use of social media platforms such as Facebook to post unsolicited advertising on personal Facebook pages of people newly diagnosed with serious diseases and conditions (Petersen et al., 2017).

Patients who are considering undertaking treatments may not be persuaded by advertising alone, a fact evidently recognized by some clinics which endeavour to represent themselves in a positive light. Other factors, such as confidence and trust in the providers' competence and the treatment's safety, developed through initial interactions or third-party recommendations, as well as national stereotypes, are also likely to play a role in decisions.

Brand Germany

When our friend came over … he said about the X-Cell Center, "Oh it's the most pristine place. It's so clean you could eat off the floor …"

(Audrey, who travelled to Germany
for SCT for Multiple Sclerosis)

As this quotation illustrates, X-Cell was seen as a destination of choice for many stem cell travellers due to its national branding which promised unique cutting-edge technology and expertise, as well as first-class service and care. Many of the Australians we interviewed were attracted to X-Cell due to the association of Germany with premium standards in both manufacturing and medical care. In this respect, German products and services were considered advanced compared to other parts of the world, including 'at home'. English proficiency, cleanliness, hygiene, and customer service were also all cited as reasons for X-Cell being a destination of choice for stem cell travellers, as the following accounts illustrate:

> And we thought, "Oh okay, that sounds a little bit more promising than India or China." There was no way I wanted to go to any of those places. Completely scared off. But something in Germany suddenly sounded, "Oh okay, it's in Europe, it's got to be, have some legitimacy and it's got to be good." So that was something that we decided to look into once I got home from hospital.
>
> (Sean, who contemplated travelling to
> Germany for SCT for spinal cord injury)

> And there's also that thing about Australia being a bit behind the times or a bit conservative when it comes to this sort of thing so I thought Germany, for some reason I had this thing in my head that, well, they make really good cars and stuff. Maybe they do know what they're doing ... I had a bit of faith in them because of their, well just Europe, you know. Just thinking that they're a bit more advanced than we are over here. I sometimes think of Australia as a bit of a backwater ... not wanting to do these sort of treatments without ... I don't know. Maybe just being overly cautious.
>
> (Audrey, who travelled to Germany for
> SCT for Multiple Sclerosis)

A distinguishing feature of the X-Cell business operation was its personalized service to patients. This did not just involve personalized interactions in the lead up to treatment but involved representatives of the clinic meeting patients at the airport or train station and accompanying them to the clinic, as well as providing transfers between their accommodation and the clinic. For some of our participants this premium service was attractive, lent legitimacy to the clinic's operations and served to cultivate trust in the treating doctors and treatments, which began when contact with the clinic was first made. The following account from Donna, whose partner died of a rare neurological disease six months after receiving treatment at X-Cell, is worth quoting at length as it offers insight into the interactions between staff and patients when patients were deciding whether or not to travel to X-Cell for treatment, including statements about the possible benefits of treatment, as well as the treatment process itself:

> They [the staff at X-Cell] were wonderful, absolutely wonderful, they phoned [and my husband] could use his computer to contact [them], and oh they were truly wonderful ... . [T]hey were more than happy to

answer any questions that we needed to know …. When they gave us the details [of treatment], they didn't pretend it was a cure, they said that it could and so we went …. And it wouldn't be forever. Oh we knew there was a chance for him to speak, that it would give him a better life, and a longer life if it took. They didn't promise that he'd be able to speak but that could have come right … .

They met us at the airport. We got there, [my husband] started the treatment on the Monday and so we got there on the Saturday and Sunday, a nurse came and took blood and then on Monday we went to the clinic, they picked us up and took us back. Tuesday was a rest day. Wednesday the purified blood was put back into him and they had to stay in the hospital for three hours lying flat on his back. Then … because of the long flight home … we flew home on the Friday.

(Donna, who travelled to Germany with her husband
for SCT for a rare terminal neurological condition)

In addition to attentive personalized care, the treatment regime was also an attractive option for many of our participants due to the short time frame that was required to undergo SCTs as well as the type of cell, which was heavily marketed by the clinic as 'safe' and more 'natural', given that the cells came from a patient's own body, a subject to which we now turn.

Stem cell magic: natural, safe, autologous

Patients or their carers, especially if they are the parents of an ill child, are likely to pursue various options before proceeding with claimed stem cell 'treatments', including the use of complementary and alternative therapies such as physiotherapy (Petersen et al., 2013). For many, however, autologous cellular products that are marketed as 'stem cell treatments' seem to have a mystique that proves especially alluring. Moreover, if the cells are taken from the patient's own body these tend to be perceived as a 'safe' option. Such popular cultural conceptions of stem cells and their magical quality are evident in some patients' and carers' accounts. For the following patients, the perceived low risk of using such autologous cells evidently figured in their reasoning for choosing X-Cell:

I wasn't interested. Not interested at all, especially in the Asian ones … because they're not your own stem cells. 'Cause they're … embryonic, yeah. But mine were my own and that's probably what sold it to me. [Because] well I didn't want anybody else's blood in me. God, I've got a body trying to recover anyway; I'm not going to risk disease and whatever.

(Rebecca, who travelled to Germany
for SCT for Multiple Sclerosis)

I came across the one in Germany and that's the one I ended up going with. Yeah. So my decision for that was because they used your own stem cells, and so I was relatively safe, provided their hygiene levels were good. [The treatment] would be relatively safe.

> (Catherine, who travelled to Germany
> for SCT for spinal cord injury)

Popular cultural narratives regarding the regenerative capacity of stem cells, along with optimistic expectations attached to the power of autologous treatments, were evident in respondents' accounts, as can be seen in the following comments:

> Well the [stem cells] just sort of move through the body and go to where the dead ones are. That's how I understand it. And just replace with new ones ... . [And] things just keep improving ... it's only two years ago and there's just little things all the time. And I say to my husband, "You're not going to believe what I did today." And that's the stem cells, and they reckon it's five years I think they keep working. They keep working into where they've got to repair. So it's pretty exciting.
>
> (Rebecca, who travelled to Germany
> for SCT for Multiple Sclerosis)

> Ivan: Autologous stem cells were extracted from his hip. Yes, bone marrow and then reinjected a few days later through lumbar puncture ... . And that was the one day, we felt it definitely made him more alert because before then he sort of wasn't tracking objects properly, wasn't engaged in what was going on around [him]. ... And within a three- to six-month period, look, that could have just happened naturally but we [Vlasta: I don't believe that] yeah, we have a strong feeling there's a sense of – [Vlasta: that it slightly accelerated his physical improvement].
>
> (Ivan and Vlasta, who travelled to Germany
> for SCT for Cerebral Palsy for their child)

While regulatory ambiguity, slick marketing, and popular cultural conceptions contribute to a flourishing market, any number of factors, singly or in combination, may predispose it to fail. This includes, significantly, regulatory intervention as well as media reporting of adverse events such as the death of a patient – as happened in the case of X-Cell – which can draw attention to otherwise unacknowledged health and safety issues and the potentially grave implications of regulatory ambiguity.

The fall of X-Cell

As noted earlier, X-Cell flourished in a context of uncertainty. However, a series of events preceded regulatory action, legal action, and media attention – events that included the complications associated with the treatment of two young boys who are believed to have had stem cells injected into their brains. Such incidents attracted scrutiny of the clinics' practices.

In 2010, the media reported that a ten-year-old boy from Azerbaijan 'almost died' following intracranial SCT for Cerebral Palsy at X-Cell (Mendick and Palmer, 2010). Another tragedy was reported three months later, after an eighteen-month-old boy died following a highly invasive intervention that involved stem cells being injected into his brain stem.[4] In Europe and around the world, the media reported on the scandal with headlines such as 'Baby death scandal at stem cell clinic' (Mendick and Palmer, 2010). The confluence of these tragedies, the criminal investigation, and the raised public awareness of potentially exploitative practices associated with the commercial stem cell industry prompted German regulators to intervene in the practice of unproven 'stem cell treatments'.

The German regulators acted in the interests of public safety but also sought innovative ways to shut down X-Cell in order to create trust in the regulatory system (Petersen et al., 2017). Due to the alignment of the EU ATMP and German Drug Law regulation, German regulators sought to establish that products marketed as 'stem cell treatments' should be considered medicinal products that take into account not only the source of cells but how they were processed in the laboratory and how they were anticipated to act upon administration to the patient. Such an approach was resisted by some in the German medical community as it was thought that by defining SCT as an ATMP it would impose new barriers and stifle innovative new treatments. Based on the aligned EU-German ATMP regulation, German regulators told X-Cell that they would require product authorization for their SCTs. This decision resulted in the closing of X-Cell in Germany.[5] Despite the controversy and the delay in acting, regulators felt that the case was a useful one, as Sasha explained:

> I think it [the X-Cell case] brought to very wide attention the risks of stem cell treatments quite nicely. Not nicely to the patient who died, but to the knowledge of the community that these are not random treatments that can be given to anyone by anyone.
>
> (Sasha, EU regulator)

Examination of the circumstances surrounding the fall of X-Cell in Germany helps to highlight how regulatory systems can be used to help engender public trust and allow questionable practices to flourish, or, as was ultimately the case, be exercised to curb non-evidence-based practices.

Conclusion

Important lessons can be learnt from the example of X-Cell and the examination of the sociocultural and politico-economic contexts that have enabled self-styled stem cell 'entrepreneurs' to operate, and that have shaped patients' perceptions of SCTs and the clinics that offer them. Currently, it is the combination of propitious regulatory contexts, such as that described for X-Cell, as well as public perceptions of high national standards of regulation and medical care along with the regenerative potential of stem cells, that are creating the conditions for the biggest businesses in the current SCT market to flourish. However, to date, lessons from this case have been largely neglected in responses to 'stem cell tourism'.

Of particular note is the exponential growth in the number of commercial clinics selling autologous SCTs in Australia and the United States (Munsie and Pera, 2014; Turner and Knoepfler, 2016). The X-Cell business model is being successfully translated in these countries and the SCT industry is booming. Both countries are publicly perceived and self-promoted as world leaders both in the development of 'cutting-edge' biotechnology and innovation, as well as in the development and enforcement of stringent regulatory standards in science and medical care. This is in spite of the concerning reality that, as in the X-Cell example, existing regulatory ambiguities and/ or loopholes exist and are being exploited with respect to, for example: clear definitions of, and standards in, the characteristics of cells that are used in treatments; the enforcement of advertising standards; the terminology used to describe cells and/or medical 'products' in policy and legislation; which governing body should act to develop more effective regulation and/or be responsible for enforcement of existing standards; the extent of harm and risk people are being exposed to; and how to define and enforce boundaries between ethical practice which ensures compassionate access to treatments whilst enabling biomedical innovation and curbing exploitative practices.

In spite of operating in areas of regulatory failure and uncertainty, commercial clinics in the United States and Australia use marketing strategies that exploit national branding in order to lend legitimacy to their operations – as was the case with 'brand Germany' for X-Cell. This culturally-specific 'branding' continues to be mobilized in direct-to-consumer marketing via websites, in clinic materials that are circulated in correspondence with potential patients, as well as via the use of social media platforms where unsolicited advertisements are posted. The presentation of autologous SCT as a safe pioneering innovation is further bolstered by the public perception that cells isolated from one's own body (autologous) are more 'natural' than those deriving from an external source (allogeneic) and therefore carry no or lower risk compared to other sources of stem cells or other interventions. The vast majority of commercial clinics now use this rhetoric as an integral part of their marketing – just as X-Cell had before them.

As with X-Cell in Germany, a concerning effect of the public image of stringent regulation and advanced medical care is public belief in Australia and the United States that available treatments meet safety and quality standards where in fact these treatments are being offered outside conventional medical practice and clinical research frameworks. For example, three women in the United States lost their vision due to complications after receiving injections of 'stem cells' obtained from their own abdominal adipose tissue (fat) for macular degeneration (Kuriyan et al., 2017). Another US woman developed fragments of bone in her face as a result of a 'stem cell facelift' using cells obtained from her adipose tissue (Jabr, 2012). In Australia, a woman died of complications following a similar autologous adipose-derived SCT for dementia (Lysaght et al., 2017). However, unlike the eventual response by German authorities to X-Cell, and despite these instances of harm, local regulators have been slow to act and curb these concerning practices.

It has also been widely acknowledged that representations of stem cell research, especially in mainstream media, fuel popular belief in the regenerative power of stem cells, depicting them as a panacea for the host of illnesses and diseases for which current medicine offers little hope for improvement or cure (Master and Resnick, 2013). There have been calls for better communication by research scientists since the beginning of stem cell research (e.g. Caulfield et al., 2016) – especially with respect to the media's portrayal of their findings – to counter stem cell 'hype' and the popular cultural imagery that contributes to public misperceptions about the unequivocal power and safety of stem cells. When updating their guidelines on standards expected in stem cell research and the development of new clinical therapies, the International Society for Stem Cell Research included a section specifically on communication and the research communities' obligation to 'provide an accurate, balanced and responsible representation of stem cell research' with failure to do so having 'tangible impacts on the expectations of the general public, patient communities, physicians and on the setting of health and science policies' (ISSCR, 2016). Whilst a laudable directive, how this is achieved in practice given the various pressures and investments of those involved, for example of journalists to attract readers and of scientists to attract funding, is yet to be seen.

Given the complex sociocultural and politico-economic dynamics at play, possible responses to the growth of 'stem cell tourism' are neither simple nor straightforward. With an absence of enforceable international regulation governing this space (Sipp et al., 2017), there is also the question of the efficacy of a national regulatory 'Whack-A-Mole' approach as businesses simply move across borders when local regulators act, as was also the case in the example of X-Cell.[3] However, the example of X-Cell suggests that for all those with a stake in stem cells – whether individuals or families considering treatment, clinicians, scientists, or others invested in improving health – finding ways to effectively communicate across these groups is essential. In particular, there is a critical need to enhance understanding and clarity concerning alternative, potentially more effective ways to address hyped claims of stem cell 'breakthroughs' in the media and associated popular cultural beliefs. At the same time, misleading advertising claims and branding that leverages national stereotypes, especially regarding standards of medical care and regulation, should be challenged. Finally, regulatory ambiguities and inadequate local enforcement that enable exploitative businesses to flourish need to be addressed.

Notes

1 The term 'stem cell tourism' is problematic and misrepresents the experiences of people who undergo these journeys. This is due to the suggestion that stem cell

tourism is comparable to other forms of medical tourism, where people travel to destinations for both leisure and to access proven medical applications, or dental/plastic surgery procedures at a cheaper rate than is available in their own country. We adopt the term here, and elsewhere, because of its common use in scientific, policy, and regulatory communities, as well the media.

2 Data drawn on in this chapter was collected as part of a larger Australian Research Council-funded study entitled, 'High hopes, high risk? A sociological study of stem cell tourism' (DP120100921).

3 The director of the X-Cell Center reportedly moved its operations to Beirut, Lebanon following closure of the German clinics.

4 It is unclear whether the cause of death was due to the type or number of cells that were used in the procedure, the method of delivery, or the follow-up care of the young boy. How many patients were treated with this method of delivery is also unknown.

5 To the best of our knowledge the X-Cell Center never applied for authorization perhaps due to a lack of clinical evidence for the SCTs they were selling.

References

Adriance, S. (2014). Fighting for the "Right to Try" Unapproved Drugs: Law as Persuasion. *Yale Law Journal Forum*, [online] Volume 124, p. 148. Available at: www.yalelawjournal.org/forum/right-to-try-unapproved-drugs [Accessed 13 June 2017].

Berger, I., Ahmed, A., Bansal, A., Kapoor, T., Sipp, D. and Rasko, J. (2016). Global distribution of businesses marketing stem cell-based interventions. *Cell Stem Cell*, 19(2), pp. 158–162.

Caulfield, T., Sipp, D., Murry, C., Daley, G. and Kimmelman, J. (2016). Scientific community: Confronting stem cell hype. *Science*, 352(6287), pp. 776–777.

Connolly, R., O'Brien, T. and Flaherty, G. (2014). Stem Cell tourism – A web-based analysis of clinical services available to international travelers. *Travel Medicine and Infectious Disease*, 12(6 Pt. B), pp. 695–701.

Copelan, E. (2006). Hematopoietic stem-cell transplantation. *New England Journal of Medicine*, 354(17), pp. 1813–1826.

Daley, G. (2012). The promise and perils of stem cell therapeutics. *Cell Stem Cell*, 10(6), pp. 740–749.

Ikonomou, L., Freishtat, R., Wagner, D., Panoskaltsis-Mortari, A. and Weiss, D. (2016). The global emergence of unregulated stem cell treatments for respiratory diseases: Professional societies need to act. *Annals of the American Thoracic Society*, 13(8), pp. 1205–1207.

ISSCR: International Society for Stem Cell Research. (2016). *Guidelines for Stem Cell Research and Clinical Translation*, [online] Available at: www.isscr.org/docs/default-source/guidelines/isscr-guidelines-for-stem-cell-research-and-clinical-translation.pdf?sfvrsn=2 [Accessed 9 June 2019].

Jabr, F. (2012). In the Flesh: The Embedded Dangers of Untested Stem Cell Cosmetics. *Scientific American*, [online] Available at: www.scientificamerican.com/article/stem-cell-cosmetics/ [Accessed 6 June 2017].

Kuriyan, A., Albini, T., Townsend, J., Rodriguez, M., Pandya, H., Leonard, R., et al. (2017). Vision loss after intravitreal injection of autologous 'Stem Cells' for AMD. *New England Journal of Medicine*, 376, pp. 1047–1053.

Lau, D., Ogbogu, U., Taylor, B., Stafinski, T., Menon, D. and Caulfield, T. (2008). Stem cell clinics online: The direct-to-consumer portrayal of stem cell medicine. *Cell Stem Cell*, 3(6), pp. 591–594.

Lysaght, T., Lipworth, W., Hendl, T., Kerridge, I., Lee, T. L., Munsie, M., et al. (2017). The deadly business of an unregulated global stem cell industry. *Journal of Medical Ethics*, [online] Available at: http://jme.bmj.com/content/early/2017/03/29/medethics-2016-104046.info/ [Accessed 13 June 2017].

Master, Z. and Resnik, D. (2013). Hype and public trust in science. *Journal of Science and Engineering Ethics*, 19(2), pp. 321–335.

McLean, A., Stewart, C. and Kerridge, I. (2015). Untested, unproven, and unethical: The promotion and provision of autologous stem cell therapies in Australia. *Journal of Stem Cell Research and Therapy*, 6(1), pp. 1–8.

Mendick, R. and Palmer, A. (2010). *Baby Death Scandal at a Stem Cell Clinic which Treats Hundreds of British Patients a Year.* [online] The Telegraph, [online]. Available at: www.telegraph.co.uk/news/worldnews/europe/germany/8082935/Baby-death-scandal-at-stem-cell-clinic-which-treats-hundreds-of-British-patients-a-year.html [Accessed 13 June 2017].

Montlake, S. (2006). Stem-Cell Tourism. *South China Morning Post*, [online] Available at: www.scmp.com/article/550313/stem-cell-tourism [Accessed 21 April 2017].

Munsie, M. and Pera, M. (2014). Regulatory loophole enables unproven autologous cell therapies to thrive in Australia. *Stem Cells and Development*, 23(1), pp. 34–38.

Ogbogu, U., Rachul, C. and Caulfield, T. (2013). Reassessing direct-to-consumer portrayals of unproven stem cell therapies: Is it getting better? *Regenerative Medicine*, 8(3), pp. 361–369.

Petersen, A., Munsie, M., Tanner, C., MacGregor, C. and Brophy, J. (2017). *Stem Cell Tourism and the Political Economy of Hope*. Basingstoke, UK: Palgrave Macmillan.

Petersen, A. and Seear, K. (2011). Technologies of hope: Techniques of the online advertising of stem cell treatments. *New Genetics and Society*, 30(4), pp. 329–346.

Petersen, A., Seear, K. and Munsie, M. (2013). Therapeutic journeys: The hopeful travails of stem cell tourists. *Sociology of Health and Illness*, 36(5), pp. 670–685.

Schnitger, A. (2014). *The Hospital Exemption: A Regulatory Option for Unauthorised Advanced Therapy Medicinal Products*. Master's thesis. Germany: University of Bonn.

Sheldon, T. (2007). The Netherlands bans private stem cell therapy. *The British Medical Journal*, 334(7583), p. 12.

Sipp, D., Caulfield, T., Kaye, J., Barfoot, J., Blackburn, C., Chan, S., et al. (2017). Marketing of unproven stem cell-based interventions: A call to action. *Science Translational Medicine*, 9(397), eaag0426.

Trounson, A. and McDonald, C. (2015). Stem cell therapies in clinical trials: progress and challenges. *Cell Stem Cell*, 17(1), pp. 11–22.

Turner, L. and Knoepfler, P. (2016). Selling stem cells in the USA: Assessing the direct-to-consumer industry. *Cell Stem Cell*, 19(2), pp. 154–157.

X-Cell Center (2007). *The Center for Regenerative Medicine*, [online] Available at: http://web.archive.org/web/20070510031852; http://xcell-center.de:80/index.php?id=%7C&L=1 [Accessed 24 Apr. 2017].

X-Cell Center (2010). *Stem Cell Treatment Clinic Specializing in Adult Stem Cell Therapy*, [online] Available at: http://web.archive.org/web/20101223082616; http://xcell-center.com/ [Accessed 6 June 2017].

3   The bioeconomies of stem cell research

Lorenzo Beltrame

Abstract

This chapter explores the implication of stem cell research bioeconomies. It starts by defining the notion of bioeconomy and the main ways in which value is produced in the commercial exploitation of biomedicine and biotechnologies – commodification of biological materials, assetization of knowledge and capacities of biotech enterprises, exploitation of the labour of tissue donors. It then analyses how value is generated in existing configurations of stem cell biobanking through two paradigmatic cases. In the first, cord blood banking, I explore how commodification and assetization are produced in both the public and the private sector, and their hybrid configurations. In the second, the economy of hESC lines circulation, I analyse how national regulations (in the USA, the UK, and Germany) about hESC research shape the emerging market, influencing the dynamics of commodification and assetization, and the creation of dominant market positions. Finally, I critically discuss tissue donor involvement, showing that the emerging market configurations entail ethical and social issues – related to exploitation of donors and the unfair distribution or sharing of the resulting benefits and economic profits – that are at the very core of the stem cell bioeconomy.

Introduction

Innovation forms the backbone of the knowledge-based economy and stem cell research represents a substantial opportunity for future innovation in the life sciences. … To ensure that the UK remains one of the global leaders in stem cell research, the UK Stem Cell Initiative (UKSCI) was established …

(UK Stem Cell Initiative, 2005, 5)

The debate on stem cell research has long been dominated by the discussion on the moral status of the human embryo – which is destroyed for deriving pluripotent human embryonic stem cell (hESC) lines. This discussion has overshadowed an entire range of concerns, problems, and issues in stem cell research and clinical application (Hauskeller, 2005), including those emerging from the economic organization of this field of biomedicine.

As the above quoted excerpt of the UK Stem Cell Initiative *Report & Recommendations* clearly shows, stem cell research is part of a global knowledge-based economy, usually named *bioeconomy*. According to the Organization for Economic Co-operation and Development (OECD), bioeconomy refers to 'the aggregate set of economic operations ... that use the latent value incumbent in biological products and processes to capture new growth and welfare benefits for citizens and nations' (OECD, 2006, 1). These 'new growth and welfare benefits' would be achieved through the generation of product markets (ibid.). In other words, bioeconomy is identified as a knowledge-based economy rotating around the commercial application of the life sciences, biomedicine, and biotechnologies.

Understanding what kind of bioeconomy is in place in stem cell research, how value is produced, how this value is exploited, and how its production is organized enables the exploration of a whole range of societal, political, and ethical implications and concerns involved in the concrete stem cell bioeconomies. In 'The bioeconomy' I will first define the notion of bioeconomy and the core concepts employed by Science and Technology Studies (STS) to analyse the economic dimension of current biosciences. Second, I will discuss the three main forms of value production detected by STS and feminist scholars in the field of bioeconomy. Finally, in 'The bioeconomy', I will explore the predominant institutional and organizational arrangements in stem cell bioeconomy with their tensions, overlaps, and 'hybrid configurations' (Hauskeller and Beltrame, 2016a). With this grid of analytical concepts, I can analyse the concrete economies in which stem cell research is carried out and the variable and complex configurations of value production and exploitation that blur some conventionally drawn boundaries between: (a) the public and the private sector, (b) the market and redistributive economy, and (c) within the market, different forms of capitalization and valuation.

In order to explore these configurations in the case of the economies of umbilical cord blood (UCB), stem cell banking, and the global circulation of hESC lines ('Redistribution and the market in UCB stem cell banking' and 'Economies of hESC lines' circulation'), I must first focus on the three main institutional sites governing the production and circulation of stem cells (adult and/or hESC) – viz. stem cell factories, biobanks, and stem cell registries. In 'Factories, banking and registries: key nodes in the stem cell bioeconomy', therefore, I will discuss the role of these institutional sites in the stem cell bioeconomy and, in particular, how regulations on stem cell production and distribution shape the emerging bioeconomic configurations.

'Redistribution and the market in UCB stem cell banking' and 'Economies of hESC lines' circulation' are devoted to the analysis of these configurations in the case of UCB stem cell banking and the circulation of hESC lines. In 'Stem cell procurement and reproductive labour', finally, I will discuss the feminist critique of the exploitation of donors involved in stem cell procurement and the related notion of *reproductive labour*.

In this chapter, I will argue that what is at stake in the so-called bioeconomy is not the peculiar mode of production in itself, but rather how it implies the intertwining of bioeconomic configurations and emerging ethical and political issues, the tensions between the production of economic *value* and the involved moral and social *values*. Indeed, the production of value in the stem cell bioeconomies is entailed in ethical problems related to issues such as the access to therapies, the sharing of economic benefit arising from the commercialization of biomaterials and technologies, and, above all, the exploitation of the embodied labour of women involved as tissue donors in stem cell procurement. As I will show, these ethical issues are at the core of the bioeconomic mode of production and they are not dealt with by the dominant discourse on the moral status of the human embryo and the related regulations of stem cell research.

The bioeconomy

Bioeconomy in policy documents is predominantly defined as the commercial exploitation of biotechnology and biomedical innovation (e.g. European Commission, 2002; OECD, 2006). STS have identified this emerging bioeconomy with a particular economic form: the capitalist market economy. According to Helmreich (2008), for example, the term *biocapital* has been introduced to give a sense of the increasing insertion of 'the substances and promises of biological materials, particularly stem cells and genomes ... into projects of product-making and profit-seeking' (p. 464). This novel form of *capital* 'fixes attention on the dynamics of labour and commoditization that characterize the making and marketing of such entities as industrial and pharmaceutical bioproducts' (ibid.). Sunder Rajan (2006, 4) has stressed how contemporary biosciences have been moving 'toward more corporate forms and contexts of research', and Cooper (2008, 19) has added that 'life becomes, literally, annexed within capitalist processes of accumulation'.

A key concept in this STS literature is that of *biovalue* introduced by Catherine Waldby (2002) which refers to 'the yield of vitality produced by the biotechnical reformulation of living processes' (p. 310). In her view, biotechnology capitalizes on life by increasing or changing the productivity of living processes at the molecular and cellular level. Biovalue is thus at the core of the process of biocapital accumulation (Rose, 2007, 6–7), and this peculiar form of capital is generated through the extraction of vitality from life itself and its transformation 'into commodities, markets, and strategies for accumulating wealth' (Franklin and Lock, 2003, 8). Adapting the Marxian formula Money-Commodity-Money' (where ' indicates the surplus value gained in exchange) Helmreich (2008) has suggested that this literature identifies the biocapitalist process as:

> B–C–B', where B stands for biomaterial, C for its fashioning into a commodity through laboratory and legal instruments, and B' for the biocapital produced at the end of this process, with ' the value added through the instrumentalization of the initial biomaterial ...
>
> (p. 472)

Other STS scholars (Birch and Tyfield, 2012; Birch, 2012, 2017) accuse this literature of relying too much on the process of commodification and of having confused the locus of value production. First, they point out that 'there is no "biovalue" in a political-economic sense' (Birch and Tyfield, 2012, 309): the market demand for health, vitality, and well-being is prompted by 'social or ethical values', while the economic value is instead created through the (knowledge) labour that produces services and products. They thus conclude that, in strictly political-economic terms, 'there is nothing intrinsically "bio-" about the value relationship itself' (p. 308), which is instead a conventional *valuation process* that realizes value in market exchanges after the transformative capacity of labour.

Birch and Tyfield also criticize the predominant focus on the materiality of biomaterials and on the centrality of commodity production posited by this STS literature that introduces the notion of biovalue. According to Birch (2017), most life science firms are valued not because they have marketed products or services but because the bio-technologies and bio-knowledge held by firms represent assets that are capitalized through financial investment practices. Value, therefore, neither resides in intrinsic aspects of biological materiality, nor is it simply produced through commodification, but it is mainly realized through the triadic process of valuation made of *assetization, capitalization*, and *financialization* (Birch, 2017, 464–470). *Assets* are resources that generate recurring earnings, since an asset 'is a tangible or intangible resource that can be used to produce value and, at the same time, has value as property' (Birch and Tyfield, 2012, 302). *Capitalization* is the process of valuation that connects the present value of an asset to the future stream of earnings through a discount rate (Birch, 2017, 466). *Financialization*, finally, is a process of profit accumulation working primarily through financial channels rather than through trade and commodity production (ibid., 464). In this sense, knowledge labour produces value not simply by creating commodities; on the contrary, knowledge is mainly monetized through intellectual property rights (IPRs), licensing, partnerships, royalties, and so on (i.e., asset-based income) (Birch, 2017, 465). Moreover, value is realized in asset-based market exchanges through the 'trading of shares or investments in the firm (i.e., financial assets) or intellectual property (e.g., knowledge assets)' (Birch and Tyfield, 2012, 311).

Finally, STS and feminist scholars have highlighted another, problematic dimension of biovalue production – namely what is called *biolabour*: for example, the appropriation of the 'reproductive' and 'clinical' labour of tissue donors (Dickenson, 2001; Waldby and Cooper, 2008) or the bodily labour of clinical trial participants in developing countries (Sunder Rajan, 2006). Accordingly, people who donate tissues, who provide bio-information, or who are recruited in clinical trials and biomedical experimental procedures are performing a form of bodily-embedded labour which is exploited and 'largely unrecognized' (Waldby, 2008, 27). This is the first step in the value chain of the biocapitalist mode of production, which is 'a form of extraction that involves isolating and mobilising the primary reproductive agency of specific body parts' (Franklin and Lock, 2003, 8). This literature has thrown light on a field of bioethical problems usually overshadowed by the dominant concern about the dignity and the moral status of the human embryo. It has, indeed, highlighted ethical and social issues related to the economy of stem cell research, and, in particular, how people donating cells, tissues, or bio-information are not always aware that their 'donation' is not necessarily inserted in a gift-relationship or in an altruistic redistribution for the common good. On the contrary, donated biomaterials are increasingly channelled into a global value chain where they are converted into commodities or assets producing profits (Gupta, 2011). This literature has also highlighted that people providing bioresources, or performing the basic biolabour, are often from developing countries in a condition of subordination to clinicians (Glasner, 2005; Gupta, 2011) and/or are economically disenfranchised (Waldby, 2008; Waldby and Cooper, 2008) as well as being excluded from the profit generated from their biological material and biolabour.

However, from a purely political economy perspective the definition of biolabour is problematic. It is not clear, indeed, how much the provision of bioresources could be considered 'labour' – which, according to the labour theory of value, is the transformative capacity converting natural resources into commodities – or whether it is instead a form of expropriation of raw material that recalls Marx's *expropriation* of the soil, which is only the basis of primitive accumulation and not the result of the capitalist mode of production (see Marx, 1977, 667–669). In any case, the biocapitalist process of accumulation starts with the appropriation of bioresources that will then enter in complex and articulated forms of value production. These forms can take the shape of both market exchanges of biomaterials and bioproducts and finance-oriented strategies extracting rent and value from knowledge-assets through IPRs regimes and from the trading of shares.

The emergent bioeconomy also features 'mobile' terrains of 'conflict and cooperation' between private commercial companies and publicly funded health and research institutions (Sunder Rajan, 2003, 88). In the field of stem cell science, there is indeed a growing commercial sector dealing with drug screening, cord blood banking and expansion, bone and skin regeneration, and stem cell mobilization agents (Gottweis et al., 2009). At the same time, therapies and clinical trials using somatic (or adult) stem cells are carried out in clinics and public hospitals, under the control of national medical regulatory bodies (Wilson-Kovacs et al., 2010). Sometimes the tension between a redistributive economy – working in the logic of the welfare state where the public sector provides health services – and an emerging market economy explodes in outright struggles between the groups of actors involved: public healthcare practitioners, university researchers, bioethicists versus private clinics, and commercial stem cell companies (Salter et al., 2015).

Besides these terrains of conflict, Gottweis et al. (2009) have stressed how one of the main distinctive traits of the stem cell economy is represented by the partnership between the public and the private sector. The role of the state is thus not only that of providing healthcare services and/or of regulating research, clinical application, and an emerging market of biomedical products, but that of financing stem cell research. In this case its role is read through the lens of the 'competition state' or the 'developmental state', in which government interventions are aimed at increasing economic competitiveness by promoting and fostering particular innovative sectors with large global markets (e.g., Salter and Salter, 2010). Public investments and international partnerships between governments, private industries, and universities are aimed at attracting venture capital and stock market investments into the field in order to develop commercializable stem cell products and therapies 'that would help shore up the nation's claim as a strong competitor among knowledge economies' (Gottweis et al., 2009, 17).

In what follows, I will analyse how these different forms of value production – commodification, assetization, and the exploitation of the embodied labour of tissue donors – are at work in two fields of stem cell bioeconomies (UCB biobanking and the circulation of hESC lines). I will also investigate how they work in the entanglement of institutional arrangements (the public and the private sector) and economic regimes (the market and the redistributive economy), and, finally, how regulations exert economic effects by shaping market mechanisms and how these bioeconomies trigger ethical and social issues.

Factories, banking, and registries: key nodes in the stem cell bioeconomy

The three main institutional sites governing the production and circulation of stem cells (adult and/or hESC) are stem cell factories, biobanks, and stem cell registries.

Stem cell factories are laboratories that isolate stem cells from several tissues – e.g., cord blood, bone marrow, adipose tissue, amniotic fluid, embryos, and so on. In cell factories stem cells are not only isolated but also processed, characterized, and engineered. From cell factories, stem cells flow to public research institutions or hospitals and transplant centres, as well as to commercial biotech companies or to clinics offering stem cell therapies. Stem cell factories can be commercial enterprises or branches of public research or healthcare institutions. Therefore, they are located at the intersection between the public and the private sector, and they operate both for the public redistributive economy and for the emerging market of bioeconomic enterprises.

As key nodes in these economies, stem cell factories are subject to regulation. At the European level, the derivation of stem cells and their use in clinical practice or their marketization as medicinal products are ruled according to an EU Directive. Accordingly, European and national states' authorities and the European Medicine Agency draw up protocols, guidelines, and principles of good manufacturing practice, good clinical practice, and good distribution practice. In the USA, the commercialization of stem cell-based products is strictly regulated by the federal US Food and Drug Administration which imposes standards of good manufacturing practices applied to sourcing, process, and pre-clinical evaluation (Salter and Salter, 2010, 95). In other countries the whole field of stem cell line creation and clinical application is less regulated, which enables both the provision (and selling) of unproved stem cell therapies and the outsourcing of clinical trials in non-orthodox settings (Sleeboom-Faulkner and Patra, 2011; see Chapter 2, in this volume). These examples indicate that lax regulatory regimes and economic conditions allow for the development of lucrative bioeconomies exploiting particular local conditions. In this sense, market mechanisms can exert different and contradictory effects. On the one hand, countries or enterprises which would participate in the global stem cell economy as suppliers of biomaterials have to enact compliance with protocols, standards, and regulations set by Western countries or international organizations (see Hauskeller, 2005). In market competition, thus, the normalizing function of scientific standards (including also guidelines on the ethics of stem cell procurement) turns into a normative one. On the other hand, however, as mentioned above, market mechanisms and profit-seeking can also foster the outsourcing of cell procurement, clinical trials, and therapy provision in countries where regulatory loopholes allow the commercialization of unproven therapies sold to ill people often travelling from Western countries – a phenomenon called 'stem cell tourism' (Ryan et al., 2010).

The second key locus of stem cell bioeconomies are biobanks. They are collections of biological materials combined with information (personal, medical, genealogical, etc.) that enable research and clinical application by providing samples and bio-information. Biobanks are also a site of biovalue production: biomaterials and bio-information can be sold and exchanged as commodities, and profitable knowledge can be extracted from the donor populations and translated into commercializable medical products or technologies (Mitchell and Waldby, 2010). Therefore, the ways in which biobanks are organized and governed are relevant to the kinds of bioeconomies that emerge. Gottweis and Lauss (2011, 66) have identified three main models: (1) entrepreneurial biobanking, carried out by commercial companies or in a public-private partnership; (2) biosocial biobanking, created, promoted, and/or funded by patient groups; and (3) public biobanking, established and funded by the state or by non-profit organizations. This typology does not exhaust all possible configurations, nor are there exclusive links between these types of biobanking and forms of value production. For example, in the case of cord blood stem cell banking, some private banks, while selling a service of personal stem cell banking, also run programmes of public donation (i.e., they provide UCB for the public healthcare system). Similarly, some public biobanks do not limit their activity to the provision of tissues or cells to public research institutions or to the public healthcare system, but they are also involved in the commercial exploitation of the banked biomaterials through, for example, public-private partnerships. As crucial nodes in the network of (public and private) actors in current biomedicine, biobanks are at the crossroads between different bioeconomic configurations that involve different opposing, and in some cases, overlapping forms of stem cells commodification – resulting in what Hauskeller and Beltrame (2016a) have called *hybrid bioeconomies.*

Finally, the third key institutional site is represented by stem cell registries. These are databases of available stem cell samples and stem cell lines, documenting the provenance of the biomaterials and relevant characterizations for their use (scientific or clinical). In the hESC economy, the European Human Embryonic Stem Cell Registry or the US National Institute of Health (NIH) Human Embryonic Stem Cell Registry plays a crucial role. First, they promote harmonization and standardization in stem cell processing and banking – through quality and safety controls and/or cell factories and biobank accreditation. Secondly, they operate as market makers: by listing hESC lines eligible for use in publicly funded research (by the EU or by the NIH), they shape the flow of funding and investment, and thus structure (or consolidate) hierarchies and dominant positions in the market among cell factories and biobanks.

As key economic actors and market makers, factories, biobanks, and registries are the main sites to study and analyse for understanding the bioeconomy of stem cell research. Regulations, institutional arrangements, and organizational modes of functioning indeed shape the resulting bioeconomic configurations and the related forms of value production, and their societal and ethical implications.

Redistribution and the market in UCB stem cell banking

UCB is a source of haematopoietic stem cells that are collected and stored in biobanks for their possible use in the treatment of haematological and oncological malignancies. The banking of UCB has developed along two different lines: (a) public biobanks that collect donated UCB and redistribute it for medical needs; and (b) a sector of commercial biobanks that sell to mothers and/or parents a service of personal stem cell banking. The public system is described as a redistributive economy – based on the logic of the solidarity of the welfare state – while the private sector is seen as a market economy of biomedical services in which individuals negotiate their own health and that of their family members (Waldby and Mitchell, 2006). This dominant view in the bioethics and biomedical literature is, however, simplistic. The ways in which biovalue is produced in UCB banking arrangements show the complexity of current stem cell bioeconomies.

In the case of public banking, while several public UCB banks operate exclusively in the redistribution of this tissue for medical needs, others are involved in regenerative medicine research on UCB stem cells. As noted by Hauskeller and Beltrame (2016a, 2016b), these latter banks have multiplex relationships with biomedical industries operating in the market: they sometimes sell discarded UCB samples to biotech companies and university spin-offs that intend to capitalize on their research, or they are in research partnership with pharmaceutical companies (which develop and commercialize the outcome of basic research on UCB stem cells). Moreover, UCB units for transplantation are exchanged internationally at a price that exceeds the costs of collection, processing, and storage (Brown et al., 2011). While it is true that prices are not generated in the classical supply-demand mechanism, and the income is used to cover the expenses of managing public banks, this quasi-market of UCB units in fact commodifies this tissue. Public banks realize a value both from exchanging UCB units considered to be of good quality and sufficient volume for successful engraftment and, in some cases, from discarded units sold for research purposes. The case of public UCB banking shows some interesting overlaps: the *use value* of UCB as a public resource in a redistributive economy is entangled with its *exchange value* as a commodity in market bioeconomies. Public banks are thus acting as a node between different regimes of biovalue production.

In the private banking sector, capitalization does not rely on commodification processes, because UCB is not sold or exchanged but owned as a biological asset (Waldby and Mitchell, 2006). The value does not reside in UCB as such, but in the technological service enabling simultaneously the *assetization* of UCB as a resource that may have a use value in the future, and the proprietary control over it. This future is both that of a possible family use of UCB as transplantable tissue in established haematological applications, and that of the development of regenerative medicine, where UCB and placenta-derived stem cells may be used in organ repair (Martin et al., 2008). Therefore, private UCB banks capitalize on commodifying a biomedical service (Hauskeller and Beltrame, 2016b). In turn, this commodification of a service of proprietary control lies not only in the assetization of UCB as a personal or family holding, but also in the assetization of the bank's technological capability and reputation. In order to attract customers, and in the competition for market share, private banks highlight their commitment to cutting-edge research in this field, as well as the accreditation by international professional organizations that assures that the bank follows best standards in UCB banking (Hauskeller and Beltrame, 2016a). In other words, accreditation certifies a set of technological and professional capabilities that are transformed into assets and then capitalized upon in the selling of a service.

The boundaries between the public and the private and, moreover, between redistribution and the market are blurred by the emergence of the so-called *hybrid banking models*. This is an umbrella term covering different banking practices that include private banks offering donation programs, specially dedicated UCB storage free to families in need of a transplant and, also, public banks that increase their income by selling private family banking options (O'Connor et al., 2012). As Hauskeller and Beltrame (2016b) have shown, hybrid banking practices do not simply mean the surpassing of the distinction between the public sector and the market, since private commercial entities operate in the public redistribution of UCB. Indeed, by participating in the international redistribution of UCB units, private banks can capitalize on export prices that exceed the cost of the storage service – i.e., if a donated UCB is sold, the revenue is higher than the profit from private storage. As Hauskeller and Beltrame (2016b, 241) conclude, partaking in the redistributive UCB economy means that the de-commodification of banking services turns into the commodification of UCB as material. Paradoxically, private banks realize value through commodification by participating in the public redistributive economy.

The case of UCB banking testifies what Sunder Rajan (2003, 87) has called the shifting relations between market commodification and public commons. Indeed, it shows that different forms of value production (commodification versus assetization) and economic regimes (redistribution versus the market) not only coexist, but in some cases crisscross the public-private boundary and evolve in hybrid modes of capitalization.

Economies of hESC lines' circulation

The use of human embryos for stem cell research is regulated differently across nations, with some countries explicitly forbidding the derivation of stem cells from the human embryo and other countries supporting more permissive regulations (Walters, 2004; Jasanoff, 2005). This has created a global and complex circuit of exchange of hESC lines intersecting to form different bioeconomic configurations. The economy of hESC line circulation not only entails interesting entanglements between (a) commodification and assetization, (b) the market and a redistributive economy, and (c) the public and the private sector but also involves what STS scholars have called reproductive or clinical labour in egg and embryo procurement, with the related ethical and social issues of exploiting women without sharing the benefit arising from hESC research and medical applications. In order to illustrate this, I will consider three paradigmatic cases of national regulation in hESC research.

The first paradigmatic case is that of the USA, where, on 9 August 2001, former President George W. Bush introduced a ban on federal funding for research on hESC lines. The NIH, the federal government's leading biomedical research organization, could fund research only on hESC lines derived prior to that date from embryos created only for reproductive purposes, and donated, without any financial inducement, through an act of informed consent (NIH, 2016a). Moreover, in the USA, the 1996 Dickey-Wicker Amendment banned the use of federal funds for any experiment involving the creation and destruction of human embryos. This regulatory framework supported the development of a dichotomous research environment, where what was unethical and forbidden in the public sector could be carried out in the market-oriented private research sector.

What was the effect of this policy on the bioeconomy of hESC research? According to bioethicist Jeffrey Kahn, the US government, in the attempt to regulate the ethics of hESC research, became a market maker; its restriction created a demand to be filled by private agents on the market (Kahn cited in Waldby, 2008, 25). Indeed, since embryo procurement and hESC derivation could not be federally funded, the private sector became the main supplier of stem cells for research. In this way, the federal government eschewed regulating hESC procurement: eggs and embryos could be obtained from voluntary donations or, as in the case of the Bedford Stem Cell Research Foundation, through the payment of women providing eggs or embryos (Waldby, 2008, 26).

However, non-federally funded does not necessarily mean private. As Salter and Salter (2010, 93) have shown, states allowing hESC research – such as California, Massachusetts, New Jersey, New York, and Wisconsin – have strongly invested public money into this field. The amount has been calculated as US$230 million in the 2005–2006 period alone. The main effect of Bush's policy was that universities or research institutes applying for NIH funds had to resort to hESC lines authorized by the NIH and listed by the NIH Human Embryonic Stem Cell Registry. Being listed in this registry, therefore, meant to have a positional advantage over competing providers of hESC lines. In summary, the resulting US regulation leads to a market structure in hESC research in which: (1) hESC lines can be provided only by enterprises non-federally funded, and (2) hESC line providers listed by the NIH registry have a dominant position in the market, since research that is federally funded can only use their cell lines.

This was the case of the Wisconsin Alumni Research Foundation (WARF) – a private, non-profit organization that manages technology transfer by patenting inventions arising from University of Wisconsin research and licensing technologies to companies. WARF holds a patent covering both the method for isolating hESCs and three cell lines developed (neural, cardiomyocytes, and pancreatic islet cells) plus a patent claiming rights over mesodermal, endodermal, and ectodermal hESC lines. WARF cell lines were listed by the NIH registry as federally fundable, and this allowed WARF to establish expensive licensing conditions – which was criticized for having created a sort of innovation bottleneck (Gottweis et al., 2009, 39). McCormick et al. (2009), by comparing the material transfer agreements (MTAs) issued by WARF and its main non-NIH listed competitors, Harvard Stem Cell Institute (HSCI), discovered that WARF has dominated the US market thanks to its fundability – notwithstanding its high fees and the fact that HSCI charged only the cost of shipping. WARF, thus, realized value not only through the exchange of hESC lines (i.e., commodification) but capitalized also its IPRs and its dominant position in the market provided by the federal regulation on hESC research (i.e., assetization).

While the US hESC regulation leaves room to develop a competitive market – and structures it by creating dominant positions – in Europe regulations are broader and seek also to govern the emerging market. In the UK the derivation of hESC lines from IVF embryos is allowed but strictly regulated by the Human Fertilisation and Embryology Authority (HFEA), which authorizes the derivation case by case. An integral part of the UK policies for governing hESC research, the UK Stem Cell Bank (UKSCB) was set up in order to facilitate a global standard of hESC line creation through offering free services of storage, characterization, and distribution to researchers following their ethical protocol everywhere on the globe. The bank is a public non-commercial institution created to reinforce the leading position of the country in this biomedical field and aims to manage the use of hESC lines and their movements in and out of the country (Stephens et al., 2008). The UKSCB decrees that cell lines must not be sold for financial gain and it makes lines available to academic researchers with minimal constraints and no upfront fees, while commercial users have to pay full costs (Waldby and Mitchell, 2006). Moreover, the UKSCB exerts an ethical control on stem cell procurement and use (Stephens et al., 2008) and by working with the International Stem Cell Initiative it seeks to set global quality and characterization standards in hESC research (Webster and Eriksson, 2008).

The case of UKSCB, as Hauskeller (2004) has argued, is paradigmatic of a view of biomedical progress that combines scientific competitiveness with the compliance with (British) ethical standards (p. 520). It reflects a sort of 'paternalist imperialism' based on a 'utilitarian orientation towards economic and public welfare on a general, not just national, level' (p. 521), which the UKSCB would like to play on the international market of hESC lines. The UKSCB was designed and funded as a public sector, redistributive economy operator that seeks to control, channel, and limit the market. The intended aim of the UKSCB strategy is to prevent a competitive market not bound to ethical standards by offering quality stem cell lines to public-sector institutions at a minimal cost. The UKSCB's cheaper, ethically controlled, and up-to-date standardized hESC lines compete with those provided by market actors who capitalize on the cell lines' exchange value. This strategy may also affect assetization processes, because it is intended to reduce the economic value of hESC lines held by other market actors and their market position. However, an open question for STS research is whether, or to what extent, this strategy has been effective in shaping and limiting the market for hESC lines.

The case of Germany, in contrast, epitomizes some contradictions emerging from restrictive regulations aimed only at defending the moral status and the dignity of the human embryo. The German Embryo Protection Law (*Embryonenschutzgesetz*) of 1990 forbids the derivation of hESC lines. HESC research, thus, is possible only on imported hESC lines but is strictly regulated according to the Stem Cell Act of 2002, amended in 2010. Research on imported hESC lines has to be authorized by the Central Ethics Commission for Stem Cell Research that verifies that used cell lines have been derived before the enforcement of the Stem Cell Act, from IVF embryos no longer needed for fertilization purposes, and that donors have not received payment, compensation, or other monetary advantage. The logic of this law is to avoid the production of hESC lines for explicit use in Germany and, in general, from embryos created for research purposes. In the context of this discussion on bioeconomies and how they are affected by different nations' policies, Germany's narrow focus on the status of the human embryo, prohibiting hESC creation, and strictly limiting their importation remains without effects on processes of commodification and marketization outside of Germany. In other words, the strictly regulated, ethical German hESC research can rely on stem cell lines provided by private commercial companies that sell them through market mechanisms.

Stem cell procurement and reproductive labour

As I mentioned in 'The bioeconomy', and as is clear in the previous section, the commerce of hESC lines relies upon what feminist scholars have called *women's reproductive labour.* In general, existing regulations on egg and embryo procurement prohibit or discourage financial compensation, because it is seen as a possible undue inducement that leads to the commodification of women's bodies and as a payment potentially exploitative of economically disenfranchised women. Feminist scholars have strongly criticized these practices as exploitative in themselves, arguing that notions and legal frameworks of informed consent and voluntary donation do not guarantee the dignity of the women involved (Dickenson, 2006; Widdows, 2009). Therefore, as these feminist scholars have strongly emphasized, the dominant focus on the moral status of the human embryo does not exhaust the ethical and social problems related to stem cell research. On the contrary, relevant ethical and social issues emerge from the economic organization of stem cell procurement.

Widdows (2009) considers embryo procurement a form of labour under conditions of exploitation. Women are induced to objectify their reproductive capacity which, by being sold, becomes commodified. And attributing a use value to body parts (*objectifying* them) and then an exchange value (*commodifying* them) 'is in a sense exploitative' because it turns something that has (ethical) value in itself – 'beyond the realm of commerce' – into 'objects of trade' (Widdows, 2009, 18–19). Braun and Schultz (2012) talk about a commercialization continuum that ranges from 'a variety of crypto-commercial strategies that enable more or less monetary transaction' (p. 140) to explicit forms of embryo or oocyte brokerage (Waldby, 2008, 23). But, according to Widdows (2009), exploitation also takes place in the case of payment or other forms of non-monetary compensations (e.g., free or discounted access to IVF treatments) because exploitation objectively occurs under the standard of disparity which lies in the imbalance, extracted under conditions of subordination, 'between the labor value invested in producing an object by the worker and the price the worker is paid for [her] labor' (Widdows, 2009, 17). In general, women providing eggs or embryos are in a relation of subordination and reliance on clinicians, they are often in a situation of poverty and economic need, and in most cases they do not have a share in either the scientific or medical benefit of hESC research nor the resulting economic profits. Gupta (2011) has shown that in poor countries the desire of a child, the stigma of infertility, and the high cost of IVF in relation to average income leave women and couples in a position of subordination thereby overpassing other considerations regarding the use and ownership of unused embryos. According to these feminist scholars, legal frameworks based on informed consent, free choice, and altruistic donation are considered inadequate since they do not solve the problem that women providing eggs neither directly benefit from stem cell research, nor have they a fair share in the financial benefits generated through it (Ballantyne and de Lacey, 2008; Dickenson and Alkorta Idiakez, 2008).

As I argued in 'The bioeconomy', the notion of biolabour could be problematic from a classical political economy point of view. These practices of tissue procurement resemble more the appropriation, or expropriation, of resources than a conventional form of labour. Accordingly, Glasner (2005, 359) argues that the production of value in this bioeconomy starts with 'a process of expropriation – i.e., dispossession – of value'. What makes this expropriation valuable is not the biovalue embedded in women's reproductive capacity itself, but the fact that it represents, especially in the global South or in pockets of poverty in the global North, a low-cost, often unregulated, 'readily available, abundant and ethically neutral supply' (ibid.) of biomaterials. Therefore, the paradox of the bioeconomy of hESC line circulation is that regulations concerned with ethics turn to produce an unfair 'moral economy' starting from practices of expropriation.

Conclusions

The bioeconomy of stem cell research and clinical application can take many forms. Value, or biovalue, does not reside in the biomaterial itself but is realized both through the technical and knowledge labour transforming the 'bio' into commodities, and through other forms of capitalization. Patents covering cell lines, methods, and techniques for isolating, growing, manipulating, and manufacturing stem cells represent assets that are capitalized in financialized-rentier regimes operating in share and stock markets. The case of stem cell banking illustrates furthermore that value can be realized by commodifying services exploiting future health promises and expectations of advances in biomedicine. If this bioeconomy is a mixture of capitalization processes based both on the production and exchange of commodities and on financialized-rentier regimes of accumulation, it also shows a complex intertwining between the public and the private sectors, and between redistribution and the market.

The case of the economy of hESC lines' circulation illustrates how the state is not simply a regulator, or a supplier of health services, but intentionally or unintentionally a market maker. Regulations can create a demand fulfilled mainly by the private sector – or by start-ups established by public institutions and in public-private partnerships. It can also generate hierarchies and dominant positions (i.e., the case of WARF hESC lines and its patents). In other cases, the state can try to govern the market through public institutions that, in a logic of redistributive economy, act as key players through particular price-making strategies, rules of distribution, ethical controls, and standard-setting for quality assessment and characterization. Sometimes attempts to limit the market are turned into devices for enabling assetization and profit-making – as exemplified in the case of UCB banking accreditation.

Finally, and most importantly, regulations appear unable to solve issues of commodification and exploitation of what some scholars call women's reproductive labour. In specific socio-economic circumstances, embryo procurement – but in general tissue procurement – relies on forms of unfair and unbalanced relationships of production that expropriate biological resources that are later valued in international markets.

What is really at stake in this bioeconomy is not the extraction of 'the latent value incumbent in biological products and processes' (OECD, 2006, 1), or, in other words, it is not just the novelty of the material means of production that is at stake but rather the resulting relations of production. What STS scholars call *biovalue* is a new class of economic value created through labour and economic processes that respond to a demand for health, fostered by ethical and social values. But in responding to this socially and ethically inspired valuation of health and well-being, the emerging bioeconomy entails economic processes of valuation that create ethical and social issues related to the exploitation of women, expropriation of biomaterials, and the unfair distribution or sharing of the resulting benefits and economic profits.

Acknowledgements

This chapter was written as part of the REGUCB project (Regulating Umbilical Cord Blood Biobanking in Europe). The REGUCB project has received funding from the European Union's Horizon 2020 research and innovation programme under the Marie Skłodowska-Curie grant agreement No. 657361.

References

Ballantyne, A. and de Lacey, S. (2008). Wanted-egg donors for research: A research ethics approach to donor recruitment and compensation. *International Journal of Feminist Approaches to Bioethics*, 1(2), pp. 145–164.

Birch, K. (2012). Knowledge place and power: Geographies of value in the bioeconomy. *New Genetics & Society*, 31(2), pp. 183–201.

Birch, K. (2017). Rethinking value in the bioeconomy: Finance, assetization, and the management of value. *Science, Technology, & Human Values*, 42(3), pp. 460–490.

Birch, K. and Tyfield, D. (2012). Theorizing bioeconomy: Biovalue, biocapital, bioeconomics or... what? *Science, Technology, & Human Values*, 38(3), pp. 299–327.

Braun, K. and Schultz, S. (2012). Oöcytes for research: Inspecting the commercialization continuum. *New Genetics and Society*, 31(2), pp. 135–157.

Brown, N., Machin, L. and McLeod, D. (2011). Immunitary bioeconomy: The economisation of life in the international cord blood market. *Social Science & Medicine*, 72(7), pp. 1115–1122.

Cooper, M. (2008). *Life as Surplus. Biotechnology and Capitalism in the Neoliberal Era*. Seattle: University of Washington Press.

Dickenson, D. L. (2001). Property and Women's alienation from their own reproductive labour. *Bioethics*, 15(3), pp. 205–217.

Dickenson, D. L. (2006). The lady vanishes: What's missing from the stem cell debate. *Journal of Bioethical Inquiry*, 3(1–2), pp. 43–54.

Dickenson, D. L. and Alkorta Idiakez, I. (2008). Ova donation for stem cell research: An international perspective. *International Journal of Feminist Approaches to Bioethics*, 1(2), pp. 125–144.

European Commission (2002). *Life Sciences and Biotechnologies. A Strategy for Europe*. Luxembourg: Office for Official Publications of the European Communities.

Franklin, S. and Lock, M. (2003). Animation and cessation: The remaking of life and death. In: S. Franklin and M. Lock, eds., *Remaking Life and Death: Toward an Anthropology of the Biosciences*. Santa Fe: SAR Press, pp. 3–22.

Glasner, P. (2005). Banking on immortality? Exploring the stem cell supply chain from embryo to therapeutic application. *Current Sociology*, 53(2), pp. 355–366.

Gottweis, H. and Lauss, G. (2011). Biobank governance: Heterogeneous modes of ordering and democratization. *Journal of Community Genetics*, 3(2), pp. 61–72.

Gottweis, H., Salter, B. and Waldby, C. (2009). *The Global Politics of Human Embryonic Stem Cell Science*. London: Palgrave MacMillan.

Gupta, J. A. (2011). Exploring appropriation of "Surplus" ova and embryos in Indian IVF clinics. *New Genetics and Society*, 30(2), pp. 167–180.

Hauskeller, C. (2004). How traditions of ethical reasoning and institutional processes shape stem cell research in britain. *Journal of Medicine and Philosophy*, 29(5), pp. 509–532.

Hauskeller, C. (2005). Introduction. In: W. Bender, C. Hauskeller and A. Manzei, eds., *Crossing Borders. Cultural, Religious, and Political Differences Concerning Stem Cell Research. A Global Approach*. Münster: Agenda Verlag, pp. 9–24.

Hauskeller, C. and Beltrame, L. (2016a). The hybrid bioeconomy of umbilical cord blood banking. *Biosocieties*, 11(4), pp. 415–434.

Hauskeller, C. and Beltrame, L. (2016b). Hybrid practices in cord blood banking. Rethinking the commodification of human tissues in the bioeconomy. *New Genetics and Society*, 35(3), pp. 228–245.

Helmreich, S. (2008). Species of biocapital. *Science as Culture*, 17(4), pp. 463–478.

Jasanoff, S. (2005). *Designs on Nature. Science and Democracy in Europe and the United States*. Princeton: Princeton University Press.

Martin, P., Brown, N. and Turner, A. (2008). Capitalizing hope: The commercial development of umbilical cord blood stem cell banking. *New Genetics and Society*, 27(2), pp. 127–143.

Marx, K. (1977 [1887]). *Capital. A Critique of Political Economy. Volume I* (trans. S. Moore and E. Aveling). Moscow: Progress Publishers.

McCormick, J. B., Owen-Smith, J. and Scott, C. T. (2009). Distribution of human embryonic stem cell lines: Who, when, and where. *Stem Cell*, 4(2), pp. 107–110.

Mitchell, R. and Waldby, C. (2010). National biobanks: Clinical labor, risk production, and the creation of biovalue. *Science, Technology, & Human Values*, 35(3), pp. 330–355.

National Institute of Health. (2016). Human Embryonic Stem Cell Policy Under Former President Bush (9 Aug. 2001–9 Mar. 2009). [online] Available at: https://stemcells.nih.gov/research/registry/eligibilitycriteria.htm [Accessed 10 June 2019].

O'Connor, M. A. C., Samuel, G., Jorden, C. F. C. and Kerridge, I. H. (2012). Umbilical cord blood banking: Beyond the public-private divide. *Journal of Law and Medicine*, 19(3), pp. 512–516.

Organisation for Economic Co-operation and Development (2006). *The Bioeconomy to 2030: Designing a Policy Agenda*. Paris: Organisation for Economic Co-operation and Development.

Rose, N. (2007). *The Politics of Life Itself. Biomedicine, Power, and Subjectivity in the Twenty-First Century*. Princeton: Princeton University Press.

Ryan, K. A., Sanders, A. N., Wang, D. D. and Levine, A. D. (2010). Tracking the rise of stem cell tourism. *Regenerative Medicine*, 5(1), pp. 27–33.

Salter, B. and Salter, C. (2010). Governing innovation in the biomedicine knowledge economy: Stem cell science in the USA. *Science and Public Policy*, 37(2), pp. 87–100.

Salter, B., Zhou, Y. and Datta, S. (2015). Hegemony in the marketplace of biomedical innovation: Consumer demand and stem cell science. *Social Science & Medicine*, 131, pp. 156–163.

Sleeboom-Faulkner, M. and Patra, P. K. (2011). Experimental stem cell therapy: Biohierarchies and bionetworking in Japan and India. *Social Studies of Science*, 41(5), pp. 645–666.

Stephens, N., Atkinson, P. and Glasner, P. (2008). The UK stem cell bank: Securing the past, validating the present, protecting the future. *Science as Culture*, 17(1), pp. 43–56.

Sunder Rajan, K. (2003). Genomic capital: Public cultures and market logics of corporate biotechnology. *Science as Culture*, 12(1), pp. 87–121.

Sunder Rajan, K. (2006). *Biocapital. The Constitution of Postgenomic Life*. Durham: Duke University Press.

UK Stem Cell Initiative (2005). *Report and Recommendations*. [online] London: Department of Health. Available at: www.ft.dk/samling/20061/almdel/uvt/bilag/245/394375.pdf [Accessed 10 June 2019].

Waldby, C. (2002). Stem cells, Tissue cultures and the production of biovalue. *Health*, 6(3), pp. 305–323.

Waldby, C. (2008). Oocyte markets: Women's reproductive work in embryonic stem cell research. *New Genetics and Society*, 27(1), pp. 19–31.

Waldby, C. and Cooper, M. (2008). The biopolitics of reproduction. Post-fordist biotechnology and women's clinical labour. *Australian Feminist Studies*, 23(55), pp. 57–73.

Waldby, C. and Mitchell, R. (2006). *Tissue Economies. Blood, Organs, and Cell Lines in Late Capitalism*. Durham, London: Duke University Press.

Walters, L. (2004). Human embryonic stem cell research: An intercultural perspective. *Kennedy Institute of Ethics Journal*, 14(1), pp. 3–38.

Webster, A. and Eriksson, L. (2008). Governance-by-standards in the field of stem cells: Managing uncertainty in the world of "Basic Innovation". *New Genetics and Society*, 27(2), pp. 99–111.

Widdows, H. (2009). Border disputes across bodies: Exploitation in trafficking for prostitution and egg sale for stem cell research. *International Journal of Feminist Approaches to Bioethics*, 2(1), pp. 5–24.

Wilson-Kovacs, D. M., Weber, S. and Hauskeller, C. (2010). Stem cell clinical trials for cardiac repair: Regulations as practical accomplishment. *Sociology of Health and Illness*, 32(1), pp. 89–105.

4 Science, ethics, and patents

Ethically-motivated barriers to the patenting of the results of human embryonic stem cell research

Fruzsina Molnár-Gábor

Abstract

Interventions into human life through the technological configuration of its basic entities such as the configuration of human embryonic stem cells are increasingly becoming possible. Because of the medical relevance of such interventions, the focus has now turned to their commercial exploitation and the ethical and legal boundaries thereof. This includes limitations on the patentability of the results of stem cell research that are among other things implemented through opening clauses into law. The wording of the opening clauses *ordre public* and *accepted principles of morality* is, however, not defined clearly in the legal texts so that their interpretation through case law on the member state and European Union (EU) level has gained increasing importance. The analysis of different case laws reveals findings on the role of opening clauses, particularly on their role as barriers to patentability in patent laws as well as in the broader legal system, and it also shows motivations that drive the limitation of patentability on an extra-legal level.

Introduction

The purpose of this chapter is to outline the foundations, main features, and consequences related to opening clauses and their introduction into the legal texts that regulate biopatenting. The chapter is divided into three main sections: the present *Introduction, Opening clauses in EU law and limited patentability,* and *Consequences of the interpretation of the opening clauses in EU patent law.* The chapter draws on the patenting of the results of human embryonic stem cell research as an example. The most well-known German patent related to this research, filed by Oliver Brüstle, describes the medical application of stem cell technology on patients being treated for Parkinson's disease, a patent that was applied for in December 1997 and approved in April 1999.[1] This example shows how closely the possibilities for intervention into biological entities are linked to a rapid increase in their potential for economic utilization. The latter has come to the fore due to patents which form the interface between scientific innovations and the economy.

A patent is an exclusive right granted for an invention, which is a product or a process that provides, in general, a new way of doing something, or offers a new technical solution to a problem (Kalanje). The grant of a patent by the government, generally for a limited period of time, over useful intangible intellectual output, provides the owner of such legal property the right to exclude all others from commercially benefiting from it (Gladstone Mills et al., 2017, §1:24). Patents are the most preferred intellectual property (IP) rights in relation to scientific innovations but are often used in different jurisdictions also as a collective term for other IP rights including but not limited to trade secrets, utility models, trademarks, geographical indications, or industrial designs. IP rights granted in the form of patents are thus legal rights which are based on the relevant national law. Such legal rights come into existence only when the requirements of the relevant law are met and, if required, it is granted or registered after following the prescribed procedure under that law. To acquire a patent, technical information about the invention must regularly be disclosed to the public in a patent application. This disclosure has, among other uses, the function of informing the general public on what inventions fall under a limited reinvention, a remaking, and a use – and therewith the patent holders are endowed with a temporary monopoly of commercial exploitation. Should the interests of the general public in the reinvention, remaking, and use of the patented subject matter outweigh the interest of the patent holder in the protection of its IP rights, additional rules that allow an adequate weighting of interests are often adopted (e.g. compulsory licensing, research exemptions). The interests of the general public in relation to a patent are also relevant with regard to the fact that patents could be taken out to protect a very destructive discovery from being reinvented by somebody else or be used (e.g. a dangerous and highly infectious virus). However, many industry players refuse to patent their innovations because the publication of the technical information in the patent application gives away too many details on the invention for their competitors to exploit.

In relation to granting IP rights on human interventions into small and complex biological systems such as embryonic stem cells, views represented in the literature point towards an additional role of patent laws, one which goes beyond economic functions and the protection of IP that promotes innovation through the granting of patents, a role that involves the possibility of drawing on ethico-moral convictions when deciding on the patentability and use of scientific inventions. In regulatory contexts, technology-based access to organisms has been interpreted as a new level of human control over nature. Such interventions into human life as well as new risks attached to them cause fear among the general public and are increasingly reflected in social debates. These debates generally concern the question of what is morally permitted in science, the undermining effects of biotechnological innovations of the notion of human nature and of the natural human lifespan, the blurring of the boundaries of species (Laritzen, 2005, 31), the intrinsic value of non-human life, and the effect of commercial application interests on the rights of those involved, particularly the right to health (German Ethics Council, 2014). The greater the number of medical applications that results from research, the more visible becomes the economic dimension that emerges through patents. The patenting of the results of human embryonic stem cell research has become one of the main causes of normative controversy in the field of life sciences in Europe. Legislators have been compelled to define specific regulations for patenting in the field of living nature. Limitations on the granting of patents for the results of stem cell research and ethically-motivated exclusions from patentability are increasingly becoming the focus of legal discussions in Europe. According to this, patent laws are often considered as having to function as a forum for respecting public moral assessments of biological interventions. Such an extension of legal norms through extra-legal ethical standards – that is, the ethicalization of patent law – is then prominently promoted by adding so-called opening clauses to regulatory texts that refer to *ordre public* and public morals. However, it is not clarified how these opening clauses should be interpreted substantially and procedurally, and *which actors are appropriate in holding debates on their interpretation or in influencing and unifying moral convictions and ethical approaches to be drawn on by their implementation.*

The Brüstle case is a prominent example of the role and interpretation of opening clauses in patent laws and will be discussed later in this chapter in relation to the phenomenon of the ethicalization of law with reference to patent law and stem cells, and to particular court decisions at various levels (EU versus national). Based on this, the next section will introduce the limitations on patentability of the results of this research through opening clauses in EU law. The third and last section will elaborate on the consequences of the interpretation of opening clauses against the background of the understanding of the subject matter.

Opening clauses in EU law and limited patentability

First approach to opening clauses and their appearance in patent law

On a general level, opening clauses are general clauses in legal acts that require the application of specific standards when interpreting these particular acts. Above all, these clauses are used to specify limitations on particular permissions defined in a legal act. The opening clauses *ordre public* and *accepted principles of morality* are not genuine formulas of supranational patent law. In fact, they were introduced into supranational patent law based on their application in national laws and on their application in international provisions on the conflict of laws (Rogge, 1998, 304) and they bring to the fore value standards founded in the overall legal system concerned (Wu, 2008).

According to this interpretation, the standards the two clauses refer to can be both legal and extra-legal. In relation to the fundamental principles and value standards of law, they mostly refer to human dignity and to standards of human and fundamental rights. Thus they ensure – in addition to the direct effect of these rights as limitations to a certain regulation – that human dignity and fundamental rights are indirectly embedded as barriers to permissions in the regulation under consideration. However, there are also standards referred to in opening clauses that exhibit points of intersection with human dignity as well as with human and fundamental rights, but that are often not congruent with these rights. Rather, they can imply a surplus meaning in such a way that, in the interpretation of these clauses, extra-legal principles such as ethical or social principles that go beyond the content of legal standards can be taken into account (Laimböck and Dederer, 2011, 662ff.; Spranger, 2012, 1203ff.; Taupitz, 2012, 3ff.).

On the supranational level, conditions for the patentability of living matter have to be regulated in a great number of countries with different legal systems and divergent public morals in a uniform manner. However, the permissibility of the commercial exploitation of an invention can be judged differently based on the different national legal and moral systems, including the applied definition of the subject matter. Because the opening clauses generally refer to comprehensive parts of the legal system and to the prevailing extra-legal standards present in a given society, according to the principle of subsidiarity established both as a principle of EU law (Art. 5 Treaty of the European Union, Subsidiary Protocol) and as a general principle of Public International Law (Feichtner, 2007) providing for the clarification of scope and limits of inter- and transnational governance, and of competing values and assumptions involved, a contradiction between harmonizing patentability and defining its conditions is at first glance inevitable.

Opening clauses in positive EU law and their relationship to the opening clauses in Public International Law

Opening clauses are an important incentive for limitations on the patentability of the results of embryonic stem cell research. They are provided directly both in international treaties and in the secondary law of the EU.

As for international treaties, according to Art. 27 of the Trade-Related Aspects of Intellectual Property Rights (TRIPS) Agreement and Art. 53, lit. a of the European Patent Convention, states can exclude inventions from patentability when the prevention of their commercial exploitation is necessary for the protection of public order and accepted principles of morality.[2] The TRIPS agreement also states that the protection of *ordre public* and morality includes the protection of 'human, animal or plant life or health or to avoid serious prejudice to the environment' (Art. 27, para. 2).

Legal protection of biotechnological inventions is standardized in EU law by the directive 98/44/EC (biotech directive).[3] In Art. 5, this directive prohibits the patenting of the human body at the various stages of its formation and development. In Art. 6, para. 1, it also places patenting in the life sciences under the provisions of public order and accepted principles of morality. Art. 6, para. 2, lit. a to lit. d of the directive even specifies these provisions using four examples (viz. processes for cloning human beings; processes for modifying the germ line genetic identity of human beings; uses of human embryos for industrial or commercial purposes – excluding inventions for therapeutic or diagnostic purposes which are applied to the human embryo and are useful to it; processes for modifying the genetic identity of animals which are likely to cause them suffering without any substantial medical benefit to man or animal, and also animals resulting from such processes).

The agreement establishing the World Trade Organization (WTO) was concluded by the EU and all its member states based on joint competence. As a result, national courts in member states can directly apply the TRIPS Agreement subject to conditions provided by the laws of the relevant member state (ECJ C-431/05, para. 48). However, WTO law also forms an integral part of the EU legal order (ibid., para. 31). In this context, the EU is obliged to respect WTO law when exercising its powers (Nguyn, 2010, 10). Accordingly, secondary law must be interpreted within the framework of the relevant international law (ECJ Cases C-402/05 P and C-415/05 P, para. 291). Recital 36–37 of the biotech directive refers to Art. 27 of the TRIPS Agreement and states that the principle by which inventions must be excluded from patentability if their commercial exploitation offends *ordre public* or morality must also be stressed in the directive. Furthermore, according to the Economic and Social Committee of the EU, it is essential that the EU legislative authority be quite clear that a legal patent protection system cannot in any way condone practices or processes that are deemed by law to be contrary to public policy and morality, notably as laid down by the Munich (European) Patent Convention (ESC, 3.3.2).

The wording and interpretation of the relevant articles in Public International Law point towards a narrow definition. However, no final position has been taken regarding the interpretation of the articles' opening clauses. Furthermore, there is no final position on the extent to which fundamental principles of law and, particularly, extra-legal standards can function as an indirect barrier to the granting and exploitation of patents through these clauses (see WTO, Appellate Body Reports). Additionally, the opening clauses in the biotech directive are less specified then they are in the TRIPS Agreement. Against the background of the increasingly dynamic scientific understanding of the relevant subject matters of patenting (Drews, 1993, 130; Gilbert, 2016, 4ff.) due to blurred boundaries between different developmental stages and types of stem cell that highlight the need for moral discussions and ethical assessment, it is particularly necessary to determine whether or not a genuine common European interpretation of these clauses exists and, if so, to what extent this is based on the common European understanding of fundamental rights and freedoms as well as public morals – or if there is in fact also room for member state interpretation beyond harmonized approaches (Plomer, 2012; Harmon et al., 2013).

Interpretation of the wording in the opening clauses 'public order and accepted principles of morality' in EU law and the role of EU institutions and member states in this area

The wording and initial interpretation of the articles in EU law also suggest a narrow definition originally developed by the Court of Justice of the EU (CJEU) (ECJ, Case 41/74; ECJ, Case 30/74). The exact scope of the concept of public order cannot be determined unilaterally by each member state without being subject to control by the institutions of the (then) European Community. Nevertheless, the CJEU also stated that the particular circumstances that justify recourse to this concept may vary from one country to another and from one period to another. It is therefore necessary to allow the competent national authorities 'an area of discretion' within the limits imposed by the Treaty of the European Union and the provisions adopted for its implementation.

As for the *implementation* of the public order and accepted principles of morality clauses, member states have room for interpretation so long as no common European standards exist regarding the interpretation of these clauses (Dederer, 2009, 11, 28). However, there is no unanimous opinion whether such common standards exist (Callies and Meiser, 2002, 430; Barton, 2004, 337). Nevertheless, the member states do not have any scope for interpretation when implementing the list of non-patentable inventions (as defined in the biotech directive regarding the specification of the opening clauses) into their national laws (ECJ, Case 456/03, recital 78; Plomer, 2006, 27–28; IPO, 2014).

Recital 39 of the biotech directive can be used as a guide for interpreting the meaning of the opening clauses. The recital states that public order and morality correspond in particular to both ethical and moral principles recognized in a member state, respect for which is particularly important in the field of biotechnology in view of the potential scope of inventions in this field and their inherent relationship to living matter. It also states that such ethical or moral principles supplement the standard legal examinations under patent law regardless of the technical field of the invention. Therefore, if a specific subject matter or process is missing from the list of non patentable inventions of the biotech directive, these clauses open up a wide scope for member states to derive fundamental law exclusions and ethically-motivated exclusions from patentability based on national interpretations.

Legislative and patent offices of member states

At first glance there appears to be great diversity in the way the above clauses are interpreted by the legislative and patent offices of member states. While some member states have simply implemented the biotech directive with identical wording, Germany for example,[4] other member states have exercised the possibility of an extended interpretation in their implementation of EU directives. In the French legislation, in addition to public order and accepted principles of morality, human dignity also represents a barrier to biopatenting.[5] The French wording refers to the dignity of the human person and not of human beings. The Austrian and the Italian legislators have expanded upon the four example exclusions of the biotech directive. The Austrian legislator has not only declared the use of human embryos for commercial and industrial purposes non-patentable, but has extended this to include all uses of human embryos. Furthermore, processes to produce chimeras from germ cells, totipotent cells, or cell nuclei of humans and animals, which are excluded from patentability in recital 38 of the biotech directive, are included on the list of non-patentable inventions implementing Art. 6 para. 2 of the biotech directive with the Austrian legislator following a relatively narrow implementation of the directive.[6] In the Italian law, Art. 4 deals with exceptions to patentability. Having excluded the patentability of the human body, the legislator narrowed this further by adding 'from the moment of conception'. The law also prohibits human cloning procedures regardless of their purpose, including therapeutic cloning, and also any use of human embryos and any patent of any invention that uses human embryos.[7]

In the UK prior to 2012, the patenting of stem cell research results was specified in the guidelines of the UK Patent Office which is now known as the Intellectual Property Office (Intellectual Property Office, 2014). According to the Office, this was necessary because the biotech directive was not taking the patenting of stem cell research results into consideration. The guidelines covered three areas of practice: the patenting of processes for obtaining hESC-lines, and the patenting of both totipotent and pluripotent stem cells. On the one hand, they refused to grant patents for processes aimed at deriving stem cells from human embryos and thereby refused to implement the exception of the directive that enables patenting of stem cells used for diagnostic and therapeutic purposes. On the other hand, to justify the patenting of pluripotent stem cells, the UK guidelines employed arguments that included the fact that these cells will not be able to develop into human beings. The main argument, however, was that the hopes attached to the research on such cells justify why commercial exploitation of innovations in this area does not infringe public order and accepted principles of morality (Spranger, 2009, 97, 100).

In the literature, national differences in the implementation of the biotech directive are traced back to culturally divergent views on the functions of patent law. One widespread view emphasizes the 'value-neutral' nature of patent law and says that its only function is to promote innovations through the granting of patents. It is not the role of patents to endorse individual research nor to grant permission for products developed on the basis of these discoveries to be placed on the market (Herdegen, 2000, 637). This view contradicts the specification in special laws concerning barriers of public order and accepted principles of morality established by the European law, and understands each national legal system in its entirety as a barrier to patenting.

The opposing views represented in the literature point to the new role of patent law, one which goes beyond economic functions. It demands that 'public order' and 'accepted principles of morality' clauses be included and specified in national laws (Calliess and Meiser, 2002, 428). However, these views rely only on arguments based on the interpretation of the opening clauses already included into patent law but do not justify the clauses themselves. Even if ethical debate on issues not determinable by science but in need of regulation gives reason for integrating opening clauses into law as they open up spaces for controversy and differing moral evaluations, the status of such clauses in regulations, how they can be interpreted and to what extent in a unified manner, has to be considered. What actually is patentable is then limited through the practical implementation of opening clauses by case law (Barton, 2004, 49–57).

The CJEU and its influence on national laws

The lack of a common European interpretation of the opening clauses – as well as the often-lacking specification of their application on new biotechnological processes and underlying biological entities as well as the member states' varying interpretations – has led to an increased involvement of the CJEU in the controversies regarding the barriers to the patenting of stem cell research results.

In 2004 at the German Federal Court of Justice (FCJ), the environmental protection organization Greenpeace took legal action against the granting of patents for procedures which are based on human embryonic stem cell lines or which imply their use. The lawsuit dealt with the issue of using surplus fertilized ova which had been donated after fertility treatment and stored for research purposes. The core argument of Greenpeace was that, as human embryonic stem cell lines are originally obtained from fertilized ova, the patented procedure represented a prohibited use of human embryos and was therefore a breach of public order pursuant to the German Patent Act (§2 Nr. 1 PatG [27 March 1992]). Following years of discussion, the patent was rejected in its original form according to the CJEU's interpretation of the biotech directive in October 2011 (ECJ C-34/10). The CJEU claimed that a 'human embryo' counts as one of the following: (1) any fertilized human ovum, (2) any non-fertilized human ovum into which the cell nucleus from a mature human cell has been transplanted, or (3) any non-fertilized human ovum where its division and further development have been stimulated by parthenogenesis. As for the argumentation, the CJEU stated that although the text of the biotech directive does not define a human embryo, nor does it contain any reference to national laws, it must be regarded as an autonomous concept of EU law and therefore interpreted in a uniform manner throughout the territory of the Union. The CJEU also stated that the illustrative list of inventions excluded from patentability should provide courts and patent offices with a general guide to interpret the reference to public order and morality. This list cannot presume to be exhaustive. However, the context and aim of the biotech directive show that the EU legislature intended to exclude any possibility of patentability where respect for human dignity could be affected. According to the CJEU, because of this respect for human dignity, the concept of the 'human embryo' within the context of the biotech directive must be understood in a broad sense (ibid., recital 34). Therefore, any human ovum, once fertilized, must be regarded as a 'human embryo' both in the context and for the purposes of the application of the directive, since fertilization is what initiates the process of developing into a human being. However, it is for the referring court to ascertain, in the light of scientific developments, whether a stem cell obtained from a human embryo at the blastocyst stage constitutes a 'human embryo' within the context of the directive. This means that, based on the opening clauses of the biotech directive, the CJEU included specific acts, considered offensive to human dignity, in their interpretation and thus in the exclusion from patentability processes. Moreover, the CJEU applied this interpretation of the clauses so as to establish a common definition of the 'human embryo' whilst disregarding various existing definitions in the member states.

It is interesting to see the decision the German national court came to after the CJEU's judgement. In German law, in para. 8 (1) of the *Embryonenschutzgesetz*, the human embryo also counts as any 'totipotent' cell removed from an embryo.[8] Nonetheless, the limited validity of the Brüstle patent on cells of ethically acceptable creation was upheld with the ruling of the German FCJ (BGH X ZR 58/07).[9] The court argued that a process qualifies as patentable when human embryonic stem cells can be obtained without destroying embryos in the process. Some methods of Brüstle permit the use of embryos that are incapable of commencing the process of developing into a human being. This is something that can be determined when the embryo is about six days old (compare, however, Chapter 8, in this volume). If a stem cell is derived from such an embryo then this is not classified as using or destroying an embryo, regardless of the fact that these embryos still contain some living cells that can be used in order to obtain stem cells that are required for purposes of specific research.

In the Brüstle case, the CJEU declared that any non-fertilized human ovum where division and further development had been stimulated by parthenogenesis constituted a human embryo. Surprisingly, however, the CJEU permitted the patenting of human stem cells derived from unfertilized ova in 2014 (ECJ C-364/13). The UK Intellectual Property Office originally turned down two patent applications due to arguments based on the case law of the CJEU in the Brüstle case issuing new guidelines in the form of practice notices for patenting in the field of stem cell research (UK IPO O/316/12).[10] The patent office felt that unfertilized human ova which have been stimulated to divide themselves should be treated similarly to embryos. The CJEU decided differently on this matter. They stated that an unfertilized human ovum where both division and further development have been stimulated by parthenogenesis does not constitute a 'human embryo' if, in light of current scientific knowledge, it does not, in itself, have the inherent capacity to develop into a human being. However, the competent British court was requested to once again consider whether it is possible for human beings to develop from the ova in question. Moreover, the CJEU also stated that the patent exclusion of human embryos represents only a minimum level of prohibition in the EU. Last but not least, the CJEU upheld the interpretation established in the decision on the Brüstle case that the concept of the human embryo within the context of Art. 6 para. 2 lit. c of the biotech directive must be understood in a broad sense (ECJ, C-456/03, recital 24). The UK High Court made an Order in January 2015 and a Consent Order in April 2015 that resulted in both applications being remitted by the Comptroller to the Examiner for further examination, and they were both granted patents (see Ipsum).

With respect to the scientific background, both cases illustrate how national courts may become bound to follow decisions of the CJEU based on poor technical information provided to courts and to eventually disregard moral convictions and ethical perceptions on the subject matter of the patenting in their own states, as was partially true due to the alleged capabilities of parthenotes to develop into human beings, as claimed in the Brüstle case (Brack, 2016). Although such judgements may be better adjusted to the scientific state of the art by means of subsequent referrals to the CJEU, this will require several years of further court work.

Beyond this immediate scientific issue, the examples show a broad extent to which the granting of patents can be reconciled with different evaluations of fundamental legal principles and ethical considerations in national and European legal systems, and they show how different the interpretations on both national and EU levels actually are. Even where the CJEU has left it to the national courts' discretion in terms of interpreting what current scientific knowledge has to say concerning the ability of various types of organisms to develop into human beings, its judgements have an effect on fields beyond patent law that underlie the divergent moral discretion in member states, such as the definition of human embryos (Bonadio, 2012; Harmon et al., 2013, 100). This has further influence on the relationship between the EU and its member states and raises questions on the legitimacy of this kind of decision-making by the CJEU and its relationship to the phenomena of the ethicalization of law.

Consequences of the interpretation of the opening clauses in EU patent law

The relationship between the EU and its member states

The increased involvement of the CJEU in the controversies regarding the barriers to the patenting of stem cell research results has an impact on the relationship between the EU and its member states.

Against the current backdrop of a Europeanization of the patent system, it is important to examine whether the CJEU is acting in an appropriate manner and whether and to what extent it is acceptable that the CJEU is in a position to make moral decisions, despite there being no agreed upon consensus among member states due to socio-moral diversity and various approaches to the scientific state of the art as a consequence thereof.

On the one hand, the CJEU's influence should be taken into consideration with regard to the member states' 'margin of appreciation' which is provided on a general level through its recognition in past decisions made by the European Court of Human Rights and the CJEU itself (Vo v France; Evans v UK; Netherlands v EP and Council). In the field of health law, which makes up the domestic competence of member states, member states' 'margin of appreciation' is originally attained by avoiding a definition of terms that can only be interpreted upon moral convictions, such as avoiding a definition of the term 'human embryo' in EU primary law and in the biotech directive. However, through the definition of the human embryo by the CJEU in the Brüstle case and its reaffirmation in the case 364/13, the process that is often referred to as a 'creeping mission expansion' of the EU in the field of health and its support by the CJEU cannot be overlooked. With the harmonization of the European legal and moral order, the discrepancy between a unified approach and the reference to divergent ethical and legal understandings in member states might gradually be mitigated.

On the other hand, promoting ethical deliberations by the CJEU is inherently connected to protecting economic interests in the single European market. In order to guarantee dynamic economic development, it can be argued that levelling out moral differences across member states and societies appears to be a necessary act and must be evaluated beyond bureaucratic and competency contests between the EU and its member states. In the Brüstle case, the CJEU did refer to a possible hindrance of European market interests by diverging barriers to patenting in the member states.[11] Providing certainty can have the effect of reducing obstacles to the common handling and free flow of scientific inventions as products or services.

All European member states' legal systems explicitly respect a moral status of the developing human life, but differ in how they understand it and how they put their understanding into laws and regulations. That making decisions on moral questions does not require direct relations to scientific achievements and definitions has already been noted in the Warnock Report. According to the dissent to this report, the determination of what constitutes '[t]he beginning of a person is not a question of fact but a decision made in the light of moral principles' (Warnock, 1984). As the status of a human embryo is predominantly derived from the principle of human dignity around the world, ad hoc changes to it might run counter to the position of dignity as a normative fundamental concept (Dederer, 2009). At the same time, the too-generous application of the concept of human dignity to human entities or even human cells might run counter to the protection of the dignity of living human beings, e.g. by voiding the concept and exacerbating its usage for defending the rights of born human beings. Also, in the long term, that the interplay between scientific progress, its moral evaluation, and amendments to the legal status of cells and organisms as 'human embryos' based on this cannot be precluded. Nevertheless, considering that the scientific understanding regarding the starting point and the characteristics and conditions of the development of human beings are processual, dynamic, and thus under-defined, it becomes vital that there is adequate space for the moral evaluation and ethical assessment of the status of cells and the different stages of the development of a human being. Against the backdrop of the activities of the CJEU in defining biological entities in the context of their patentability, the question arises as to *whether courts are the appropriate actors to influence and to unify moral convictions and ethical approaches in this manner.*

The *role of the judiciary* seems to become ever more central in the intersection of science, economic innovation, and patent law assessment of legal subject matters such as stem cells and how they are made. Whether this is acceptable is a key question, especially since most of the judicial decisions, because of their moral significance, rarely lead to harmonized decisions throughout single courts, not to mention across national jurisdictions. Judges' understanding within a single court can differ significantly with regard to their own role in deciding on a morally contested subject matter – and even more so across cultural, moral, and legal contexts. Judicial cases, in which questions have to be decided that are as similarly charged with moral convictions as the question of the patentability of embryonic stem cells, reveal courts' reservations in becoming involved in moral debates. In a case where the legal status of embryos had to be decided, judges of the Irish Supreme Court stated that:

> [t]his is not a forum for deciding principles of science, theology or ethics. This is a court of law which has been requested to interpret the Constitution and to make a legal decision of interpretation on an article in the Constitution.

Furthermore, judges also stated:

> I do not consider that it is for a Court of law, faced with the most divergent if most learned views in the discourses available to it from the disciplines referred to, to pronounce on the truth of when human life begins. Absent a broad consensus on that truth, it is for legislatures in the exercise of their dispositive powers to resolve such issues on the basis of policy choices.
>
> (Judge Denham, *M.R. v T.R*, at 46 and
> C. J. Murray, *M.R. v T.R*)

Other higher courts, such as the US Supreme Court, have also pointed out such difficulties related to the moral evaluation of questions surrounding human life:

> [w]e need not resolve the difficult question of when life begins, when those trained in the respective disciplines medicine, philosophy and theology are unable to arrive at any consensus, the judiciary, at this point in the development of man's knowledge, is not in a position to speculate as to the answer.
>
> (Judge Blackmun in *Roe*).

Room for ethics-based decision-making by non-legislative actors

In accordance with the examples given on the patenting of stem cells, however, courts appear to be given a considerable amount of discretion on the European level on ethically-motivated decision-making in the granting of patents. A large part of the existing legal regulations in patent law were created before new types of disputes related to biological and medical developments became apparent. Besides the significant importance of patent office practices in the process of granting patents, the courts' job to continue developing the law is of essential importance for the flexibility of patent law. Their granting of patents and decision-making usually involves weighing up both the rights and obligations and the interpretation of legal requirements in accordance with human and fundamental rights. In addition to this, the influence that factors beyond material law have on the practice of court rulings is also recognized (Görlitz and Voigt, 1985, 156) – although courts themselves regularly reject the further development of the law which is influenced by extra-legal standards. However, their decisions in individual cases can lead to an ethically-motivated configuration of the possibilities and limitations of patenting stem cell research results and can also lead to such a configuration of fields of law which lie outside of patent law (Spranger, 2012, 1203; Taupitz, 2012, 3ff.). The coordination between patent-granting practice and case law is characterized in particular by an extra-legal foundation and can in turn influence the development of legislative proposals.

Ethicalization of law and its legitimacy

Being also a system of justice, laws rely on moral convictions and ethical ideas in the broadest sense. However, the concept of direct and ethically-motivated influence on law has only been an issue in legal science debates for the past decade. This concept of ethicalization – understood as an increasing and reinforced extension of legal norms through extra-legal ethical standards (Vöneky, 2012, 68) – addresses fundamental questions concerning the relationship of ethics and the law. It considers not only the relationship between ethical standards and legal norms but makes a justification also of the inclusion of ethical norms in positive law necessary.

Against the current backdrop of a Europeanization and supra-nationalization of the patent system, ethically-motivated barriers incorporated in opening clauses have to be considered within the broader context of concepts of legitimacy. In this context, it is questionable whether a shift towards non-legislative influence on ethically-motivated barriers can be justified as a form of management for the patent system, let alone be based on market interests. Assessing this shift this chapter shows the extent to which morality has a normative link to legally protected rights and can thus transform the normative persuasiveness of ethical concepts into legal argumentation in patenting. The opening clauses contain the moral and ethical norms that are foundations of the law and are decisive in awarding moral-ethical considerations power to establish legally legitimate barriers to patenting.

However, when discussing emerging patentability questions regarding human embryos and human embryonic stem cells, the *prevention of legal uncertainty* is of the utmost importance. Many open questions pertain, regarding for example the point from which onwards and through which technologies human life can be created and recognized, or whether artificial and natural totipotency should be dealt with differently. It is also still undecided whether processes of biological development or the results of such processes of human development are the objects of laws and regulations. I have touched on these issues, but I cannot resolve them in this chapter.

I want to close with the question of who should be given the power to decide on these challenges and why. This question cannot be resolved if there is not any previously adopted viewpoint in relation to a more abstract and thus more fundamental debate that takes into account how moral questions can be handled in the world of *legal values and basic rights*. Conceptual approaches to this question could be the result of a reflexive and explicit societal debate that the law is destined to adopt (Lenoir, 2000, 1426), with due regard to (maybe scientific or evidence-based) reality. Either way it is important to strive for social adequacy and semantic specificity on decisions in a science-driven sector, where market forces easily undermine claims that fundamental rights and dignity are and ought to be foundational to the law. Silence and time-delays on public discussions regarding the moral understanding of the subject matter (Tanner, 2012) as a basis for limitations of their economic exploitation must be revisited in order to avoid both overregulation and a 'dialog of the deaf' (Caplan et al., 2005), and there is both room and need for further (international) initiatives in this respect.

Acknowledgements

Very special thanks go to Klaus Tanner and Gösta Gantner. This chapter would not have been written without the thorough discussions with them, which resulted in a joint project conducted together with further experts, funded by the German Federal Ministry of Education and Research since May 2016 (01GP1603B).

The author's special thanks go to Thomas Holstein for his generous time and effort in clarifying biological basics to a non-expert and for his openness to discuss regulatory issues in an interdisciplinary manner. Any errors related to the scientific understanding of the subject matters are those of the author.

Special thanks also go to the Heidelberg Academy of Sciences and Humanities for integrating this project into its scientific program.

Finally, special thanks also go to the editors of the book who have patiently guided the author during the improvement of this chapter.

Notes

1 Patent *Neurale Vorläuferzellen, Verfahren zu ihrer Herstellung und ihre Verwendung zur Therapie von neuralen Defekten*, DE 19756864 C1.
2 Agreement on Trade-Related Aspects of Intellectual Property Rights (World Trade Organization [WTO]) 1869 UNTS 299; Convention on the Grant of European Patents (as amended) 1065 UNTS 199, UKTS No 20 (1978), Cmd 7090.
3 Directive 98/44/EC of the European Parliament and of the Council of 6 July 1998 on the legal protection of biotechnological inventions, Official Journal L 213, 30/07/1998 P. 0013 – 0021.
4 Patent Act as published on 16 December 1980 (Federal Law Gazette 1981 I p. 1), as last amended by Article 2 of the Act of 4 October 2016 (Federal Law Gazette I p. 558).
5 Art. 17–18, Loi no 2004-800 relative à la bioéthique, August 6, 2004. Art. 17 I, Code de la propriété intellectuelle.
6 Compare Art. 2 of the Austrian Patent Act: Patentgesetz, BGBl. I Nr. 42/2005, June 9, 2005.
7 Conversione in legge, con modificazioni, del decreto-legge 10 gennaio 2006, n. 3, recante attuazione della direttiva 98/44/CE in materia di protezione giuridica delle inventioni biotechnologiche, (Act of February 22, 2006), Nr. 78, Gazzetta Ufficiale Nr. 58, 10 March 2006.
8 Act of the Protection of Embryos (The Embryo Protection Act) of 13 December 1990 (BGBl. I p. 2746), last amended by Article 1 of the Law of 21 November 2011 (BGBl. I p. 2228).
9 In April 2013, however, the Brüstle patent was revoked by the European Patent Office, https://www.epo.org/news-issues/news/2013/20130411a.html.
10 See further: http://webarchive.nationalarchives.gov.uk/20140603093549/; http://webarchive.nationalarchives.gov.uk/20140603093549 / http://www.ipo.gov.uk/pro-types/pro-patent/p-law/p-pn/p-pn-stemcells-20120517.htm. UK High Court of Justice, Chancery Division, Patents Court, Judgement of April 17, 2013, [2013] EWHC 807 (Ch), available at http://www.bailii.org/ew/cases/EWHC/Ch/2013/807.html. Reference for a preliminary ruling from High Court of Justice (Chancery Division) (United Kingdom) made on 28 June 2013 – International Stem Cell Corporation v Comptroller General of Patents (Case C-364/13), available at http://curia.europa.eu/juris/document/document.jsf?text=&docid=140398&pageIndex=0&doclang=en&mode=req&dir=&occ=first&part=1&cid=102572.
11 'The lack of a uniform definition of the concept of human embryo would create a risk of the authors of certain biotechnological inventions being tempted to seek their patentability in the Member States which have the narrowest concept of human embryo and are accordingly the most liberal as regards possible patentability, because those inventions would not be patentable in the other Member States. Such a situation would adversely affect the smooth functioning of the internal market which is the aim of the Directive'. ECJ, Fn. 40, recital 28.

References

Barton, T. (2004). *Der "Ordre Public" als Grenze der Biopatentierung. Konkretisierung und Funktion der Vorbehalte zum „ordre public" und zum menschlichen Körper in der EG-Biopatent-Richtlinie einschließlich der Umsetzung ins deutsche Recht.* Berlin: Erich Schmidt.

Bonadio, E. (2012). Biotech patents and morality after Brüstle. *European Intellectual Property Review,* 34(10), pp. 433–443.

Brack, H. P. (2016) Post Brüstle Developments in EU Biotech Patent Law at the CJEU. *Epi Information,* [online] Volume 1, pp. 20–25. Available at: http://information. patentepi.com/1-16/post-bruestle-developments.html [Accessed 1 Nov. 2017].

Calliess, C. and Meiser, C. (2002). Menschenwürde und Biotechnologie: Die EG-Biopatentrichtlinie auf dem Prüfstand des europäischen Verfassungsrechts. *Juristische Schulung,* 42(5), pp. 426–432.

Caplan, A., Fossett, J., Shickle, D., McGee, G., Ouellette, A., Carroll, K. and Bjarnadottir, D. (2005). Lessons Across the Pond: Assisted Reproductive Technologies in the United Kingdom and United States. *American Journal of Law and Medicine,* 31(4), pp. 419–446.

Dederer, H.-G. (2009). Patentierbarkeit der Forschungsergebnisse im Zusammenhang mit human-embryonalen Stammzellen, insbesondere mit dem sog. Therapeutischen Klonen – Aspekte des deutschen und europäischen Rechts. In: J. Straus, P. Ganea, and Y.-Ch. Shin, eds., *Patentschutz und Stammzellforschung. Internationale und rechtsvergleichende Aspekte.* Berlin-Heidelberg: Springer, pp. 367–398.

Drews, U. (1993). *Taschenatlas der Embryologie.* Stuttgart: Thieme.

ECJ, Case 41/74, Van Duyn v Home Office, [1974] ECR 1337, AT P. 1350. (Upheld by: ECJ 30/77, Recs 33–35).

ECJ, Case C-431/05, Merck Genericos Produtos Farmaceuticos v. Merck & Co [2005] ECR-I-7001.

ECJ, Joined Cases C-402/05 P and C-415/05 P, Yassin Abdullah Kadi v. Council and Commission, [2008] ECR-I-6351.

ECJ, Case C-456/03, Commission v Italy, [2005] ECR I-5335.

ECJ, C-34/10, Oliver Brüstle v Greenpeace eV, Judgement of 18 Oct. 2011.

ECJ, C-364/13, International Stem Cell Corporation v Comptroller General of Patents, Designs and Trade Marks, Judgement of 18 Dec. 2014.

Economic and Social Committee, Opinion on the 'Proposal for a European Parliament and Council Directive on the legal protection of biotechnological inventions', Official Journal C 295, 7 Oct. 1996, P. 0011.

ECtHR, Vo v France [2005] 40 E.H.R.R. 12; Evans v United Kingdom [2008] 48 E.H.H.R. 34; Netherlands v European Parliament and Council (C-377/98) [2001] E.C.R. I-7079.

Feichtner, I. (2007). Subsidiarity. In: R. Wolfrum, ed., *Max Planck Encyclopedia of Public International Law.* Oxford: Oxford University Press.

German Ethics Council. (2014). Stem Cell Research – New Challenges of the Ban on Cloning and Treatment of Artificially Created Germ Cells? Ad hoc Recommendation of the German Ethics Council of 15 September, 2014.

Gilbert, S. C. (2016). *Developmental Biology,* 11th ed. Sunderland, MA: Sinnauer Associates.

Gladstone Mills, J., Cress Reiley, D., Clare Highley, R. (2017). *Patent Law Fundamentals,* 2nd ed. Clark Boardman.

Görlitz, A. and Voigt, R. (1985). *Rechtspolitologie*. Opladen: Westdeutscher Verlag.

Harmon, S. H. E., Laurie, G. T. and Courtney, A. (2013). Dignity, plurality, patentability. The unfinished story of Brüstle v Greenpeace. *European Law Review*, 38(1), pp. 92–106.

Herdegen, M. (2000). Die Erforschung des Humangenoms als Herausforderung des Rechts. *JuristenZeitung*, 55, pp. 633–641.

High Court of Ireland, *M R v. T R & Ors* [15 Nov 2006] IEHC 359.

Intellectual Property Office. (27 June 2014). Statutory Guidance: Inventions Involving Human Embryonic Stem Cells, [online]. Available at: https://www.gov.uk/government/publications/inventions-involving-human-embryonic-stem-cells/inventions-involving-human-embryonic-stem-cells-27-june-2014 [Accessed 1 Nov. 2017].

IPO Patent Decision. (16 Aug. 2012). BL number O/316/12, [online]. Available at: https://www.ipo.gov.uk/p-challenge-decision-results/p-challenge-decision-results-bl?BL_Number=O/316/12 [Accessed 1 Nov. 2017].

Ipsum. Online Patent Information and Document Inspection Service, GB2431411 – Parthenogenic Activation of Human Oocytes for the Production of Human Embryonic Stem Cells, [online]. Available at: https://www.ipo.gov.uk/p-ipsum/Case/PublicationNumber/GB2431411 [Accessed 1 Nov. 2017].

Kalanje, C. M. Role of Intellectual Property in Innovation and New Product Development. *WIPO*, [online]. Available at: http://www.wipo.int/export/sites/www/sme/en/documents/pdf/ip_innovation_development.pdf [Accessed 1 Nov. 2017].

Laimböck, L. and Dederer, H.-G. (2011). Der Begriff des "Embryos" im Biopatentrecht - Anmerkungen zu den Schlussanträgen von GA Yves Bot v. 10. März 2011, Rs. C-34/10 – Brüstle – Zugleich eine Kritik des Kriteriums der" Totipotenz", 60 *Gewerblicher Rechtsschutz und Urheberrecht – Internationaler Teil*, 8–9(60), pp. 661–667.

Laritzen, P. (2005). Stem cell biotechnology and human rights: Implications for a posthuman future. *Hastings Center Report*, 35(2), pp. 25–33.

Lenoir, N. (2000). Europe confronts the embryonic stem cell research challenge. *Science*, 287(5457), pp. 1425–27.

Neurale Vorläuferzellen II [2012] X ZR 58/07, 27 November 2012 [Federal Supreme Court of Germany].

Nguyn, T. T. (2010). *Competition Law, Technology Transfer and the TRIPS Agreement*. Northampton, MA: Edward Elgar.

Online Patent Information and Document Inspection Service, GB2440333 – Synthetic Cornea from Retinal Stem Cells Derived from Human Parthenotes, [online]. Available at: www.ipo.gov.uk/p-ipsum/Case/ApplicationNumber/GB0621069.4 [Accessed 1 Nov. 2017].

Plomer, A. (2006). Stem Cell Patents: European Patent Law and Ethics Report, [online]. Available at: www.researchgate.net/publication/258448369_Stem_Cell_Patents_European_Patent_Law_and_Ethics_Report_FP6_%27Life_sciences_genomics_and_biotechnology_for_health%27_SSA [Accessed 1 Nov. 2017].

Plomer, A. (2012). After Brüstle: EU accession to the ECHR and the future of European patent law. *Queen Mary Journal of Intellectual Property*, 2(2), pp. 110–135.

Protocol on the Application of the Principles of Subsidiarity and Proportionality to the Treaty of Amsterdam (10 Nov. 1997) [1997] OJ C340/105 (Subsidiary Protocol).

Rogge, R. (1998). Patente auf genetische Informationen im Lichte der öffentlichen Ordnung und der guten Sitten. *GRUR*, 51(3–4), pp. 303–309.

Spranger, T. M. (2009). Aspekte des britischen, amerikanischen, kanadischen und australischen Rechts. In: J. Straus, P. Ganea, and Y-Ch. Shin, eds., *Patentschutz und Stammzellforschung. Internationale und rechtsvergleichende Aspekte*. Berlin-Heidelberg: Springer, pp. 97–110.

Spranger, T. M. (2012). Case C-34/10, Oliver Brüstle v. Greenpeace e.V., with annotation. *Common Market Law Review*, 49(3), pp. 1197–1210.

Supreme Court of the United States of America, *Roe v. Wade* [22 Jan. 1973] 410 US 113.

Tanner, K. (2012). Bioethik als kulturelle Praxis. In: H.-G. Kräusslich, and W. Schluchter, eds., Marsilius-Kolleg 2010/2011. Heidelberg: Winter, pp. 213–223.

Taupitz, J. (2012). Menschenwürde von Embryonen – europäisch-patentrechtlich betrachtet. Besprechung zu EuGH, Urt. v. 18 Oct. 2011 – C-34/10 – Brüstle/Greenpeace. *Gewerblicher Rechtsschutz und Urheberrecht*, 114(1), pp. 1–5.

Vöneky, S. (2012). Ethische Standards im Wissenschaftsrecht. In: V. Epping, C. Flämig, R. Grunwald, et al., eds., *Wissenschaftsrecht, Wissenschaftsrecht – Zeitschrift für deutsches und europäisches Wissenschaftsrecht*. Beiheft 21: Wissenschaft und Ethik, pp. 68–96.

Warnock, M. (1984). Report of the Committee of Inquiry into Human Fertilisation and Embryology.

WTO, Appellate Body Reports, *European Communities – Measures Prohibiting the Importation and Marketing of Seal Products*, WT/DS400/AB/R / WT/DS401/AB/R, adopted 18 June 2014, DSR 2014:I, p. 7. Appellate Body Report, *United States – Measures Affecting the Cross-Border Supply of Gambling and Betting Services*, WT/DS285/AB/R, adopted 20 Apr. 2005, DSR 2005: XII, p. 5663 (and Corr.1, DSR 2006: XII, p. 5475) Appellate Body Report, *China – Measures Affecting Trading Rights and Distribution Services for Certain Publications and Audiovisual Entertainment Products*, WT/DS363/AB/R, adopted 19 January 2010, DSR 2010: I, p. 3.

Wu, M. (2008). Free trade and the protection of public morals: An analysis of the newly emerging public morals clause doctrine. *The Yale Journal of International Law*, 33(1), pp. 215–251.

5 Current developments in human stem cell research and clinical translation

Stephanie Sontag and Martin Zenke

Abstract

This chapter provides an overview on the history of stem cell science and reviews current methods and shortcomings in the generation of pluripotent stem cells and their application in pharmaceutical and clinical settings. Pluripotent stem cells are a transient cell population during embryonic development. They are unique in (i) exhibiting an unlimited self-renewal potential *in vitro* and (ii) generating all cells and tissue types of the adult organism. Therefore, they hold great promise for regenerative medicine and patient-specific therapies. Pluripotent stem cells isolated from the human embryo, referred to as embryonic stem cells (ES cells), have raised ethical controversies and legal restrictions. The discovery that fully developed somatic cells can be reprogrammed into pluripotent stem cells, referred to as induced pluripotent stem cells (iPSCs), revolutionized stem cell research and therapies. Today, iPSCs and their derivatives are used to (i) model and study diseases in a dish, and (ii) screen for new drugs, and test their efficacy and safety. In addition, initial iPSC-derived cell replacement therapies are in phase I and II clinical trials.

Totipotent, pluripotent, and multipotent stem cells during human development

Stem cells are characterized by their unique self-renewal and differentiation potential. During the development of multicellular organisms, stem cells give rise to all specialized cell types and tissues that make up the entity of the adult organism, hence the name *stem cells*. Stem cells also self-renew, that is, they undergo a number of symmetric cell divisions thereby maintaining a pool of stem cells with the same developmental properties.

In general, totipotent, pluripotent, and multi- or oligopotent stem cells are distinguished during mammalian development. Totipotent stem cells have the potential to develop into an entire organism including embryonic and extra-embryonic (e.g. placenta) tissue (Boroviak and Nichols, 2014; De Paepe et al., 2014). In humans, only the zygote (fertilized egg) is totipotent. After fertilization the zygote divides forming cell clusters of different cell numbers (two-cell stage, four-cell stage, etc.). These blastomeres gradually lose totipotency until at day five of human embryogenesis a blastocyst has

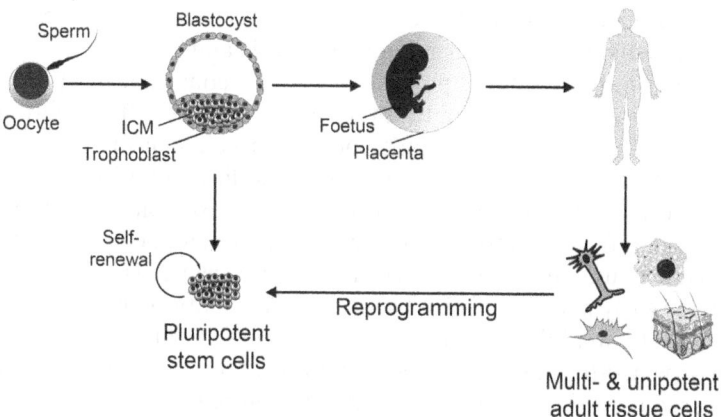

Figure 5.1 Pluripotent and multipotent stem cells during human development.

Notes: After fertilization of an oocyte, a blastocyst is formed within a week. The ICM of the blastocyst gives rise to all cell and tissue types of the foetus while the trophoblast surrounding the ICM gives rise to extraembryonic tissues, e.g. placenta. Along this development, stem cells are distinguished according to their differentiation potential. The fertilized oocyte is totipotent because it can develop into an entire organism, including embryonic and extraembryonic tissues. The ICM contains pluripotent stem cells that can still develop into all embryonic tissues and possess self-renewal potential. Isolated pluripotent stem cells from the blastocyst are referred to as embryonic stem (ES) cells. Pluripotency is lost during development when tissues and organs are specialized. Multipotent stem cells, which can differentiate into specific lineages, and terminally differentiated unipotent (somatic) cells, which have no further differentiation capacity, persist until adulthood. Such cells can be isolated and reprogrammed into pluripotent stem cells, referred to as induced pluripotent stem (iPS) cells.

formed. The blastocyst is a fluid-filled sphere that contains specific cell clusters, at this stage referred to as inner cell mass (ICM), surrounded by a single layer of epithelial cells, referred to as trophoblast. Cells of the ICM are pluripotent as they form the embryo proper and can give rise to all embryonic tissues but not to extra-embryonic tissues. In contrast, the trophoblast gives rise to extra-embryonic tissue, such as the placenta, but it does not contribute to embryonic tissues (Figure 5.1).

After the implantation of the blastocyst on days 7–9 the ICM develops into the epiblast (De Paepe et al., 2014). Epiblast stem cells are still pluripotent but as they have undergone more developmental stages they are referred to as primed pluripotent stem cells to distinguish them from naïve pluripotent stem cells of the pre-implantation ICM (Boroviak and Nichols, 2014). Epiblast stem cells lose their pluripotency when a thickened structure, called the primitive streak, is formed along the midline of the epiblast, which at this stage defines the body axes and orientations of the future embryo: cranial (head) versus caudal (feet), anterior (front) versus posterior (back), as well as left versus right end. During this process (known as gastrulation) the single-layered epiblast is reorganized into a three-layered gastrula forming the three germ layers: ectoderm (outer layer), mesoderm (middle layer), and endoderm (inner layer).

In the following weeks specific organs and tissues arise from these germ layers such as epidermal and neuronal cells from ectoderm, cardiac and hematopoietic cells from mesoderm, and pancreatic and gastrointestinal cells from endoderm. During these processes, cells are continuously specialized to fulfil their later functions and thus they gradually lose differentiation potential. Along this development, multipotent stem cells can still differentiate into a variety of specialized cell types but this differentiation capacity is restricted to a specific lineage and/or organ. For example, hematopoietic stem cells can only differentiate into erythrocytes or leukocytes but not into hepatocytes. Thus, pluripotent stem cells are a transient cell population only found in the early phases of embryogenesis, while multipotent stem cells are found throughout adulthood in various tissues and organs and are responsible for replenishing adult cell pools. Therefore, they are also referred to as adult/tissue/organ-specific stem cells. Upon terminal differentiation, those cells lose their ability to give rise to more differentiated progeny and are then called unipotent or somatic cells (Figure 5.1).

Reprogramming cell fate and the discovery of induced pluripotency

For decades it was thought that the differentiation of stem cells was irreversible. In 1893, August Weismann postulated that unnecessary genetic information must be deleted upon the commitment of cells to a specific fate and that lineage commitment is therefore permanent (Takahashi and Yamanaka, 2016). This dogma, known as the Weismann barrier, was questioned when in 1962 John Gurdon was able to demonstrate a proper tadpole development by transferring somatic nuclei of tadpole intestinal epithelium cells into enucleated oocytes (Gurdon, 1962). Such oocytes divided and gave rise to a tadpole that was genetically identical to the somatic cell. The technique itself, later known as somatic cell nuclear transfer (SCNT) or cloning, was already established by Briggs and King in 1952 when they generated viable tadpoles and frogs after transferring nuclei from early blastocysts into enucleated oocytes (Briggs and King, 1952). Still, Gurdon's report revolutionized the field of developmental biology as he could show that (i) somatic cells contain all genetic information necessary to develop an entire organism, and that (ii) the process of differentiation is reversible.

Yet Gurdon's results were heavily debated as the frequency with which normal adult frogs were generated after SCNT was very low (1%) and thus it was speculated that the frogs were the cloned offspring of contaminating intestinal stem cells rather than of fully differentiated somatic cells (Yamanaka and Blau, 2010). Moreover, Gurdon's results were based on a particular frog species (*Xenopus laevis laevis*) and could not be reproduced in other species, much less in mammals. But in 1997 Wilmut and colleagues successfully transferred sheep mammary epithelial cells into enucleated oocytes, and the cloned sheep 'Dolly' reached global publicity (Wilmut et al.,

1997). However, the idea that fully specialized somatic cells could be reprogrammed to form an entire organism, meaning they would be genetically totipotent, was not fully accepted until clonal mice were generated by SCNT from mature B cells, T cells, and olfactory neurons (Hochedlinger and Jaenisch, 2002; Eggan et al., 2004). To this day, twenty-three animal species have been cloned via SCNT including cats, dogs, mice, pigs, horses, goats, rats, and rabbits (Loi et al., 2016). Despite the technical advancements the overall efficiency of SCNT is still very low (1%–5%) and generated animals often show genetic abnormalities, aberrant gene expression, increased cancer susceptibility, and premature death most likely due to incomplete reprogramming of the somatic nucleus (Yamanaka and Blau, 2010; Loi et al., 2016).

A different approach demonstrating that cell fates are plastic and can be reprogrammed was made by cell fusion. In 1983, Blau et al. fused human amniocytes (cells isolated from the amniotic fluid surrounding the embryo) and murine muscle cells (Blau et al., 1983). These fusion products, referred to as heterokaryons, do not divide – this means that two distinct nuclei remain intact in the cytoplasm, each with a full set of chromosomes. Since the two nuclei are from different species, chromosomes can be distinguished and the impact of one genome on the other can be studied. Blau et al. found that the human amniocytes quickly activated and expressed muscle-specific genes and that this reprogramming of amniocytes is mediated through the joined cytoplasm. These findings were met with great excitement but also scepticism as they showed, in accordance with Gurdon's report, that cell fate is determined by gene expression rather than gene content. Moreover, these cell fusion experiments indicated that gene expression is regulated by trans-acting repressors or activators and that their balance determines cell fate decisions.

Around the same time, Martin Evans, Matthew Kaufmann, and Gail Martin established pluripotent cell lines derived from the ICM of pre-implantation mouse blastocysts, cells that later became known as ES cells (Evans and Kaufman, 1981; Martin, 1981). These cells proved to be pluripotent based on their self-renewal and three germ layer differentiation potential. Several years later, James Thompson and colleagues succeeded in establishing blastocyst-derived human ES cells (hESCs) (Thomson et al., 1998). Together, the fact that murine and human pluripotent cells could be maintained *in vitro* opened new possibilities in developmental biology and raised new hopes but also ethical concerns in regenerative medicine. The fusion of murine and hESCs with somatic cells demonstrated that the pluripotent state can override the somatic state meaning that the genome of the somatic cell was epigenetically reset to a pluripotent state due to the activity of gene regulators present in the ES cells. Thus, reprogramming *via* cell fusion revealed (i) that cell fate is not permanent but regulated by gene expression, and (ii) it also suggested that reprogramming factors exist in pluripotent cells that upon introduction into somatic cells can rejuvenate those to a pluripotent state.

The existence of such factors was further supported by the discovery of so-called master transcription factors regulating cell identity. In 1987, Schneuwly et al. attracted attention when they showed that the overexpression of the transcription factor *Antennapedia* at a specific larval stage in *Drosophila melanogaster* caused the formation of a second pair of legs instead of antennae (Schneuwly et al., 1987). In the same year, Weintraub and colleagues overexpressed the muscle-specific transcription factor MyoD in a fibroblast cell line and demonstrated conversion to muscle cells (Davis et al., 1987). Several years later, expression of the red cell transcription factor Gata1 converted monocyte/macrophage cells into erythroid cells (Kulessa et al., 1995). Conversely, the introduction and expression of transcription factor Pu.1 switched erythroid-megakaryocytic into monocytic cells (Nerlov and Graf, 1998). These direct fate conversions of somatic cells, a process known as transdifferentiation, supported the notion that cell differentiation and specialization are dependent on gene expression and thus can be modulated. Yet these findings were still met with caution as they were obtained with cell lines, which are generally more plastic than primary cells (Graf and Enver, 2009). These objections were dismissed when Thomas Graf and colleagues demonstrated transdifferentiation of primary mouse B cells into macrophages with Cebpα overexpression (Xie et al., 2004).

The results from SCNT, cell fusion, and transdifferentiation experiments over one century led to the conclusion that upon differentiation cells do not lose genetic information. In fact, fully differentiated somatic cells contain the complete genetic equipment as totipotent cells and the differentiation process is reversible. Thus, the Weismann barrier principle was refuted. With the knowledge that cell fate was determined by the epigenome rather than the genome and that master transcription factors were responsible for cell identity and lineage specification, scientists were motivated to identify such reprogramming factors, meaning transcription factors that would reprogram somatic cells to a pluripotent state.

In 2006, Shinya Yamanaka and his coworkers discovered that pluripotency could be induced by the ectopic expression of four transcription factors in somatic cells (Takahashi and Yamanaka, 2006). Today, these factors are globally known as canonical or Yamanaka reprogramming factors and the generated cells are referred to as *induced pluripotent stem cells* (iPSCs). Yamanaka and colleagues had started by screening mouse ES cells for pluripotency-associated and specific transcription factors. Together with reports of other groups they collected a list of twenty-four candidate factors, including Nanog, Oct4, Klf4, Sox2, c-Myc, Stat3, and β-catenin. Candidate factors were expressed in mouse embryonic fibroblasts (MEF) individually and in a twenty-four factor combination and scored for the appearance of a pluripotent phenotype. In fact, they obtained ES cell-like colonies after transduction of MEF with all twenty-four factors. Subsequently, they eliminated one factor at a time, testing twenty-four combinations of twenty-three factors each. Finally, they showed that the minimal set of four transcription factors – Oct4, Klf4, Sox2, and c-Myc – was sufficient to generate iPSCs. These cells showed ES cell characteristics with regard to their growth potential and morphology, and that they were capable of teratoma formation (that is, a tumour containing tissues of all three germ layers) upon injection into immunocompromised mice.

However, some ES cell-specific genes were rather lowly expressed in first-generation iPSCs compared to ES cells, and the promoter region of some ES cell regulators was not fully demethylated in iPSCs. Additionally, generated iPSCs contributed to foetal but not to adult chimeric mice upon blastocyst injection (2N complementation assay). Therefore, the first-generation iPSCs appeared to be only partially reprogrammed. One year later, Okita and Yamanaka et al. reported on iPSCs with ES cell-like gene expression and methylation patterns that were capable of chimera formation and even germline transmission meaning that the iPSC genome is passed on to the next generation via functional germ cells (Okita et al., 2007). Already in 2007, human fibroblasts were reprogrammed into iPSCs with the same Yamanaka factors (Takahashi et al., 2007) and also with a different factor combination (Oct4, Sox2, Nanog, Lin 28) (Yu et al., 2007).

Yamanaka's publications are regarded as milestones in the field of stem cell research. In 2012, John Gurdon and Shinya Yamanaka were awarded the Nobel Prize of Physiology and Medicine as their work has 'revolutionized our understanding of how cells and organisms develop' (Institutet, 2012). iPSCs are a powerful tool for regenerative medicine without leading to the ethical controversies that arise using hESCs.

Reprogramming systems for iPSC generation

Since the discovery of iPSCs many systems have been developed to deliver reprogramming factors to somatic cells. The choice of the reprogramming system is of utmost importance as the generated iPSCs need to be suitable for the intended downstream applications. For example, clinical application grade iPSCs and their derivatives are required to be free of exogenous reprogramming factors (transgenes). Thus, reprogramming systems are generally divided into integrative and non-integrative delivery methods depending on whether or not the transgenes are integrated into the genome of the target cells. Additionally, reprogramming systems are chosen according to their reprogramming efficiency, the simplicity with which a method can be implemented in the existing workflow, the reprogramming time, and the target tissue/cell type available.

Initially, Yamanaka and colleagues delivered the reprogramming factors individually with a retroviral infection (transduction) into murine and human fibroblasts (Takahashi and Yamanaka, 2006; Takahashi et al., 2007). Soon thereafter, somatic cells were reprogrammed with lentiviruses, which, in contrast to retroviruses, infect dividing but also non-dividing cells and thereby increase the target cell population (Yu et al., 2007; Rao and Malik, 2012).

Retroviruses and lentiviruses are known to be highly infectious, which is to say that they have a high transduction rate. Therefore, reprogramming efficiencies with these vectors are in general high. Also, retroviral and lentiviral transductions are commonly used in many laboratories and on many cell types, so that iPSCs can be obtained rather easily and cost-effectively (Rao and Malik, 2012). But as both viruses are naturally single-stranded (+) RNA viruses, they reverse transcribe their genetic information into double stranded DNA during their life cycle and integrate arbitrarily and at several positions into the host genome. Frequently, they integrate into active loci and thus they can alter both the genotype of the infected cell and potentially also the phenotype. Both retro- and lentiviral transgenes are silenced after integration, although lentiviral transgene expression is maintained longer thus resulting in a higher reprogramming efficiency. In order to temporally control transgene expression, inducible lentiviral vectors were developed to switch transgene expression on and off by doxycycline administration (Stadtfeld et al., 2008; Soldner et al., 2009). Furthermore, polycistronic lentiviral vectors were reported to increase reprogramming success as all transgenes are delivered to the target cell simultaneously (Carey et al., 2009; Sommer et al., 2009).

As the potential of iPSCs in regenerative medicine, clinical research, and drug development became more obvious, the focus shifted to the use of non-integrating reprogramming systems. Nowadays, a variety of viral (Adenovirus, Sendai virus) and non-viral (episomal plasmids, protein, mRNA, minicircle DNA, small molecules) reprogramming methods are available that generate iPSCs with zero transgene footprint (Rao and Malik, 2012). The most commonly used are Sendai viruses or episomal plasmids.

The Sendai virus is a single-stranded (-) RNA virus that, during its complete life cycle, never has a DNA stage and is therefore not able to integrate into the host genome (Fusaki et al., 2009). Indeed, Sendai virus-based reprogramming has been validated for multiple cell types and species, including fibroblasts, hematopoietic cells, keratinocytes, epithelial cells, and skeletal myoblasts (Fusaki et al., 2009; Seki et al., 2011; Trokovic et al., 2013; Tucker et al., 2013). The viral RNA genome is gradually lost after several passages of iPSCs and the absence of transgenes can readily be verified by molecular techniques such as gene-specific polymerase chain reaction (PCR). Lately, a second generation of Sendai viruses has been generated that is temperature sensitive and thus viruses can be removed by culturing iPSCs at 38°C for five days (Ban et al., 2011). Additionally, Sendai viruses reprogram somatic cells with a high efficiency (approx. 1%) and are therefore a good choice for iPSC generation (Rao and Malik, 2012). However, it has to be noted that the manual production of Sendai virus particles is technically challenging, and commercially available ready-to-use Sendai virus reprogramming vectors are expensive (Rao and Malik, 2012).

A more cost-effective alternative for generating footprint-free iPSCs are episomal plasmids, which are equipped with viral origins of replication (e.g. oriP/EBNA from Epstein-Barr virus) to ensure a sufficient expression period to reprogram somatic cells (Yu et al., 2009). While the initial reprogramming efficiencies were poor, in recent years progress has been made in plasmid design, and reprogramming efficiencies of episomal plasmids range from 0.001% to 0.02% (Rao and Malik, 2012). So far no study has reported on the integration of episomal plasmids in the host genome but as DNA could potentially integrate into the genome, whole genome sequencing should be performed on episomal reprogrammed iPSCs to ensure transgene-free iPSCs.

Mechanisms and roadblocks of reprogramming

After a decade of transcription factor-mediated reprogramming towards pluripotency, the underlying mechanisms are still poorly understood (David and Polo, 2014; Smith et al., 2016; Takahashi and Yamanaka, 2016). However, understanding the molecular events and circuitries during reprogramming is of great importance not only from the point of view of basic science but also from that of applied research. In order to use iPSCs and their derivatives for pharmaceutical and medical applications they need to be reproducibly generated in a high throughput manner, of a consistent quality that is safe for the patient. Additionally, understanding reprogramming will help to shed light on the steps of embryonic development, which consequently will help to improve *in vitro* differentiation strategies. Therefore, great efforts are being undertaken to elucidate this 'black box' of reprogramming (Cyranoski, 2014b; Tapia and Scholer, 2016).

These days, reprogramming is generally divided into three phases: initiation, maturation, and stabilization (David and Polo, 2014). In the initiation phase, the somatic cell fate is lost, meaning somatic cell genes are silenced. Simultaneously, the first pluripotency-associated genes (PAG) are expressed, including Cadherin 1 (Cdh1, E-Cadherin) and Epcam (David and Polo, 2014). These are especially important for the mesenchymal-to-epithelial transition (MET), which is characteristic for the initiation phase of reprogramming. In this transition, mesenchymal cells, e.g. fibroblasts, decrease in size, acquire an epithelial cobblestone-like morphology, and arrange into tightly packed colonies. These changes are induced by all reprogramming factors synergistically (Smith et al., 2016). Importantly, epithelial cells, e.g. keratinocytes and hepatocytes, do not need to undergo MET. Hence, they are reprogrammed much faster and more efficiently than mesenchymal cells such as fibroblasts.

Cells that have successfully overcome the first barriers will enter the maturation phase of reprogramming. Here, the first core endogenous pluripotency genes are expressed, e.g. Fbxo15, Sall4, Nanog, Esrrb, and Oct4 (David and Polo, 2014). By the end of the maturation phase also late pluripotency genes are expressed, e.g. Sox2, Dppa4, and Pecam1 (David and Polo, 2014). Particularly important in this phase is the silencing of the introduced exogenous reprogramming factors. Cells that have not activated their endogenous pluripotency network, including Oct4, Klf4, Sox2, and c-Myc, will not become stable iPSCs.

In the stabilization phase, cells have already re-acquired pluripotency and this is maintained independently of exogenous reprogramming factors. It is reported that this stabilization phase continues for several passages after initial colony emergence (Chin et al., 2009). During this time, telomeres are elongated (Marion et al., 2009), (in female cells) the X-chromosome is re-activated (Stadtfeld et al., 2008), and the epigenetic memory is reset (Kim et al., 2011). The latter refers to epigenetic marks that remain from the somatic cell (hence somatic memory) and can result in differentiation propensities towards the previous somatic lineage. Note that, while in the initiation and the maturation phase, epigenetic information is remodelled by modifications in the histone code, and in the stabilization phase re-activated DNA methyltransferases erase somatic DNA methylation signatures (David and Polo, 2014). This process can be accelerated when iPSCs are cultured with DNA methyltransferase inhibitors, e.g. 5-azacytidine.

Reprogramming is a stepwise process and many roadblocks need to be overcome in order for cells to become pluripotent. While most cells initiate reprogramming and undergo MET, only a few cells succeed in activating their endogenous pluripotency network (Smith et al., 2016). Other cells die or pause in a partially reprogrammed unstable iPSC state. Small molecules that help in overcoming roadblocks can enhance reprogramming efficiency and kinetics (e.g. TSA [trichostatin A], valpronic acid [VPA], Ascorbic acid [vitamin C], etc.) (Huangfu et al., 2008; Esteban et al., 2010; Zhu et al., 2010; Downing et al., 2013).

Molecular features and functional characteristics of iPSCs

iPSCs must be extensively characterized for pluripotency on the morphological, molecular, and functional levels. Morphology and growth characteristics are amongst the earliest selection criteria for iPSCs. Like hESCs, human iPSCs grow in tight but flat colonies with a distinct outer border and pronounced individual cell borders (Maherali and Hochedlinger, 2008; Asprer and Lakshmipathy, 2015). Each cell has an epithelial cobblestone-like morphology and prominent nucleoli. In contrast, mouse iPS and ES cell colonies are dome shaped with a 'shiny' border. Under appropriate culture conditions, stably pluripotent iPSC clones grow indefinitely as they have an extensive self-renewal potential. Frequently, human iPSCs are cultured on MEF with specific culture media. Today, MEF-free culture systems are available using fully synthetic culture media, which are particularly important for clinical purposes. Irrespectively of the culture system, human iPSCs rely on basic fibroblast growth factor (bFGF) signalling to maintain pluripotency. In contrast, mouse iPSCs and ES cells rely on the leukaemia inhibitory factor (LIF) to maintain pluripotency (Maherali and Hochedlinger, 2008).

On the molecular level, human iPS (and ES) cells are characterized by the expression of key pluripotency surface markers (Asprer and Lakshmipathy,

2015). Further to this pluripotent immunophenotype, human iPSCs show a similar gene expression profile to hESCs, including the high expression of core pluripotency genes such as Oct4, Nanog, and Sox2 (Asprer and Lakshmipathy, 2015). Today, microarrays and bioinformatic tools (e.g. PluriTest) are available to classify iPS and ES cell clones as pluripotent according to their global gene expression profile (Muller et al., 2011). Similarly, iPSCs are comparable to ES cells epigenetically, and bioinformatic tests (e.g. Epi-Pluri-Score) have been developed to classify iPSC and ES cell clones as pluripotent according to their DNA methylation profiles (Lenz et al., 2015). Importantly, fully reprogrammed iPSCs must maintain their pluripotency characteristics with their activated endogenous pluripotency factors after transgene silencing or excision.

Functionally, truly pluripotent iPSCs, such as ES cells, demonstrate three germ layer differentiation potential *in vitro* and *in vivo*. *In vitro*, this potential is assessed with spontaneous differentiation cultures (EB assays) in which iPSCs are cultured as three-dimensional aggregates, referred to as embryoid bodies (EB) (Asprer and Lakshmipathy, 2015). This way embryonic development is mimicked and differentiated cultures are analysed for the expression of germ layer specific genes. *In vivo*, the pluripotency of human iPSCs is evaluated by teratoma formation (Maherali and Hochedlinger, 2008; Asprer and Lakshmipathy, 2015). A teratoma is an encapsulated tumour containing tissues of all three germ layers. For a teratoma assay, iPSCs are subcutaneously injected into immunodeficient mice and after four to nine weeks the developed tumour is histologically examined for structures of the three germ layers. However, due to a lack of standardization and to ethical concerns regarding animal experiments, teratoma assays are no longer considered the gold standard for human iPSC characterization. Instead, gene expression and DNA methylation profiles (see above) are used to assess pluripotency (Asprer and Lakshmipathy, 2015).

Note that the reprogramming and maintenance of iPSCs can enrich (or deplete) cells with a given or acquired genetic lesion. Hence, iPSCs need to be examined periodically for genetic aberrations in their karyotype (Maherali and Hochedlinger, 2008; Asprer and Lakshmipathy, 2015). Only fully characterized, stable and pathogen-free iPSCs with a normal karyotype can be used in clinical and pharmaceutical settings.

iPSCs in stem cell research and clinical translation

Pluripotent stem cells, particularly iPSCs, offer appealing opportunities to model and investigate diseases *in vitro* (that is, in the laboratory), and to treat, alleviate, or cure diseases *in vivo* (that is, in the patient). Prior to the discovery of iPSCs, genetic disorders or diseases were mainly investigated in genetically modified animals, particularly in mice and rats (Avior et al., 2016; Trounson and DeWitt, 2016). Until recently, animal models were the most valuable tools to analyse disease-causing mechanisms and to test promising candidate drugs. However, due to the evolutionary divergence of rodents and humans in regard to development and physiology (e.g. size and organ function), animal models are limited, and findings cannot be transferred to humans directly. Furthermore, many genetic disorders (e.g. Lesch-Nyhan syndrome, Bloom syndrome) are not faithfully recapitulated in mice and thus a reliable mouse model is lacking. Further still, animal studies are connected with an ethical debate and intensive regulatory affairs, making them unsuitable for high-throughput screenings of newly developed drugs.

As an alternative, researchers have collected and isolated primary cells from patients to model diseases *in vitro*. This approach relies on the availability of a large cohort of patients and the accessibility of the affected tissue (Avior et al., 2016). While tissues – such as bone marrow, skin, or blood – can be obtained from patients through more or less invasive methods, other affected cell types, e.g. neurons in neurological disorders (such as Alzheimer's disease and Parkinson's disease), cannot be isolated from patients. Moreover, isolated primary cells from patients can only be cultured for a limited time span before they undergo apoptosis (cell death) and/or senescence (cell ageing). Additionally, a constant supply of patient material would be required to conduct consecutive experiments, which ultimately would be difficult to compare due to the different genetic backgrounds of each patient. Thus, there are several limitations and shortcomings when using primary patient materials for studying human diseases and developing new therapies.

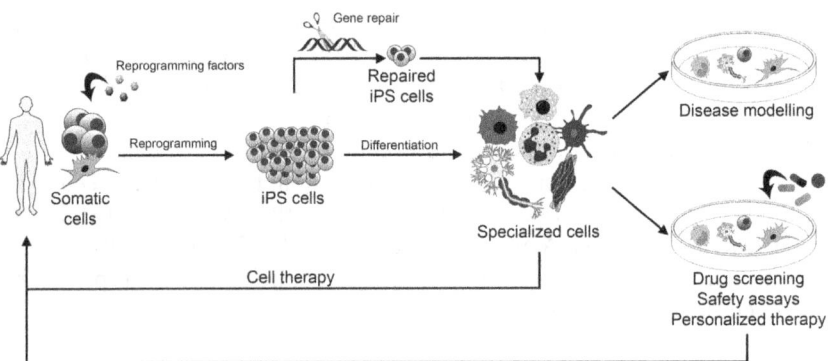

Figure 5.2 Applications of iPSCs.

Notes: Somatic cells from patients (e.g. blood or skin cells) are isolated and reprogrammed with the Yamanaka factors to generate patient- and disease-specific iPSCs. iPSCs are differentiated *in vitro* into specialized tissue cells – e.g. cardiomyocytes, neurons, or hepatocytes. This allows for the modelling of diseases in a dish, the screening for new drugs, the testing of their safety, and the evaluating of personalized therapies for patients. Additionally, differentiated patient-derived cells can be used in cell therapy and transplanted in patients to substitute for (i) trauma-associated tissue degeneration (e.g. cardiac infarction) or for (ii) disease-associated tissue degeneration (e.g. amyotrophic lateral sclerosis [ALS], diabetes type I, etc.) after the repair of disease causing mutation(s).

iPSCs bypass the limitations of animal models and primary patient samples as (i) they can be generated from a multitude of patient-derived cells, e.g. blood, skin, hair, or urine samples (more precisely renal tubular cells present in urine); (ii) they can be grown indefinitely in culture to large cell numbers; and (iii) they can be differentiated virtually into any cell type of the adult body (Avior et al., 2016; Trounson and DeWitt, 2016). Thus, they provide a continuous supply of patient- and disease-specific cells with the same genetic background that can be used for *in vitro* disease models, high-throughput drug screenings, and eventually personalized therapy and cell therapy (Figure 5.2). In cell therapy this opens the possibility of replacing diseased and degenerated tissue in patients (e.g. cardiomyocytes after cardiac infarction or neurons after spinal cord injury) with healthy cells differentiated from patient-matched iPSCs (Trounson and DeWitt, 2016). In this context, iPSCs are particularly appealing as they are considered to immunologically match the patient and in transplantations would not lead to or minimize immune rejection.

It is to be noted that because iPSCs resemble ES cells in their self-renewal and differentiation capacity, ES cells are equally useful for the aforementioned applications. Even cell therapy is nowadays conceivable with patient-specific ES cells, as Tachibana et al. reported in 2013 on the generation of hESCs via SCNT (Tachibana et al., 2013). However, neither the isolation of hESCs from blastocysts nor the generation of hESCs via SCNT comes without ethical concerns; in many countries there are legal restrictions.

Despite these obvious drawbacks, ES cells are still considered vital for stem cell research as many groups found significant differences in gene expression and DNA methylation profiles between iPSCs and ES cells (Takahashi and Yamanaka, 2016; Tapia and Scholer, 2016). Such differences are probably due to an insufficient epigenetic resetting during reprogramming meaning that iPSCs retain a memory of their previous somatic cell type. Therefore, it is disputable whether iPSCs can be considered equivalent to ES cells. Large efforts have been undertaken to compare hESCs with iPSCs and donor matched SCNT-ES cells on the molecular and functional level (Takahashi and Yamanaka, 2016). Additionally, a more detailed understanding of the reprogramming mechanisms is necessary to improve the generation of epigenetically fully reprogrammed iPSCs that resemble ES cells in all aspects. Until then, ES cells are an indispensable control in iPSC-based disease models, and the combined use of patient-specific iPSCs and ES cells is considered a crucial complement.

In addition, isogenic iPSCs have become the standard control in most iPSC-based disease models and drug screenings (Avior et al., 2016). Such isogenic iPSCs have the same genetic background but only differ in a specific alteration in the genome, e.g. iPSCs carrying a specific mutation versus the same iPSCs not carrying the mutation. This way, observed phenotypes (that is, the clinical manifestation of the disease) and possible effects of administered drugs *in vitro* can be correlated to the disease causing mutation in question rather than to other genetic alterations. However, it is well known that iPSCs and ES cells acquire mutations and even chromosomal aberrations upon long-term culture (Tapia and Scholer, 2016). Thus, isogenic iPSC or ES cell controls can only be used reliably in disease models if they are checked regularly for genetic (only involving isolated genomic loci) or chromosomal alterations (involving large changes in chromosomes).

Isogenic iPSCs or ES cells for disease models were only made largely applicable through the advent of new genetic engineering tools. Since the new millennium editing the genome of pluripotent or multipotent stem cells was achieved, e.g. with zinc finger nucleases (ZFN), but these approaches were labour- and cost-intensive, and the outcomes hard to predict (Mussolino and Cathomen, 2012; Gaj et al., 2013). In recent years, transcription activator-like effector nucleases (TALEN) and especially clustered regularly interspaced short palindromic repeats (CRISPR)-guided Cas9 nucleases have revolutionized stem cell research (Wang et al., 2016). Following the elucidation of the molecular basis of their function, nucleases could be designed to target any nucleotide sequence in any genome. Particularly, CRISPR-guided Cas9 nucleases are readily programmed to target a user-defined genomic region and the CRISPR/Cas technology can be implemented into standard laboratories with little effort and cost (Wang et al., 2016). Nowadays CRISPR/Cas nucleases are used widely throughout the stem cell field to efficiently engineer iPSCs or ES cells with required genetic mutations, deletions, or insertions. Additionally, patient-specific iPSCs or ES cells harbouring disease causing mutations or genetic alterations can be repaired *in vitro* with CRISPR/Cas nucleases, differentiated into the required and affected cell type, and might at some stage be transplanted back into the patient (Figure 5.2). Thus, iPSCs and ES cells and their derivatives do not only offer ways to screen newly discovered drugs or to personalize treatments, but they could potentially be the key to cure a plethora of genetically inherited diseases.

Hitherto, a variety of neurological (e.g. Alzheimer's disease, Parkinson's disease, amyotrophic lateral sclerosis [ALS], Down syndrome, etc.), haematological (e.g. chronic myeloid leukaemia [CML], Polycythaemia vera, primary myelofibrosis, etc.), cardiovascular (e.g. long QT syndrome type 1 and type 2, etc.), and metabolic (e.g. diabetes type 1, Lesch-Nyhan syndrome, etc.) disorders have successfully been modelled with iPSCs and ES cells (Bellin et al., 2012; Avior et al., 2016). Clinical trials for several new drugs for the aforementioned diseases are presently in phase I or II, meaning that they are being tested on healthy volunteers or affected patients to demonstrate their safety and efficacy.

Furthermore, cell replacement therapy is progressing into the clinics and in individual cases patients have already received iPSC-derived autologous (that is, from the patient itself) tissue transplants (Trounson and DeWitt, 2016). In 2014, a patient suffering from age-related macular degeneration (AMD) causing progressive blindness was transplanted with retinal pigment epithelium cells differentiated *in vitro* from iPSCs and has today almost completely re-gained eyesight (Cyranoski, 2014a; Kyodo, 2015). Until now, such examples have been rare and similar procedures – e.g. the recovery of limb movement and sensation through transplantation of spinal cord neurons, or the regeneration of heart tissue after myocardial infarction through transplantation of cardiomyocytes – were only successful in rodents and non-human primates. Despite these hurdles, such and other cell replacement therapies (e.g. transplantation of neurons for Parkinson's disease, transplantation of encapsulated insulin-producing β-islets for diabetes type I) are in phase I or II clinical trials and are expected to be tested on larger patient cohorts in the near future (Trounson and DeWitt, 2016).

However, despite this encouraging progress, there are multiple challenges ahead. Most ongoing iPSC-based clinical trials focus on neurodegenerative and monogenic (that is, caused by mutation in a single gene) disorders (Avior et al., 2016). Such diseases affect a large number of patients, particularly in an increasingly aged society, and they are relatively easy to model *in vitro*. Other more complex disorders (e.g. autism spectrum disorder, several forms of leukaemia, etc.) that involve mutations in several, possibly poorly understood genomic regions require extensive characterization of the primary patient material and the derived iPSCs following genetic engineering of several loci in order to generate isogenic control iPSCs. Thus, such disorders are hitherto under-represented in clinical trials. It has to be mentioned that drug screenings on iPSC-based disease models were mostly performed with promising candidate drugs that were already approved for other diseases (Avior et al., 2016). In such cases, typical preclinical safety assessments can be omitted and clinical trials can be started much faster. However, in the future more high-throughput screenings with newly developed drugs will be necessary in order to address as yet unmet clinical needs.

Furthermore, the establishment of iPSC repositories and registries (Moran, 2013; Seltmann et al., 2016) must be further expanded to collect and share disease-specific iPSC lines and their characteristics. It is widely agreed that reliable disease models require more than one iPSC clone per patient and ideally even several iPSC clones from several patients. Central and publicly available iPSC repositories would facilitate the use of a larger set of patient-specific iPSCs in disease models across the globe and this would ultimately strengthen the results of drug testings, cytotoxicity, and safety assays in preclinical trials (Avior et al., 2016). Consequently, more reliable and supported preclinical studies with patient- and disease-specific iPSCs would reduce the number of clinical trials, including animal experiments that need to be stopped because a drug proves to be ineffective or toxic in animals or patients.

The key to using iPSCs and their differentiated progeny for clinical applications and regenerative medicine is their extensive characterization with regard to pluripotency, differentiation capacity, genetic or chromosomal aberrations, and viral (e.g. human immunodeficiency virus [HIV], etc.) or bacterial (e.g. mycoplasm, etc.) contaminations (Takahashi and Yamanaka,

2016). The importance of such information was shown in a clinical trial for AMD in Japan, which was hampered because patient-derived iPSCs demonstrated genetic mutations and therefore raised safety issues for planned transplantations (Trounson and DeWitt, 2016). In the future, more effort needs to be invested in the thorough characterization of iPSCs, including an in-depth analysis of their genome, to prevent clinical trials from failing and to ensure the safety of patients.

While today the automated generation, characterization, and differentiation of iPSCs are not yet implemented in most research laboratories and pharmaceutical companies, such technologies will accelerate the path of iPSC-based therapies into clinics, as multiple patient-specific iPSCs can be processed in parallel with reduced costs and time efforts (Avior et al., 2016; Yandell, 2016). As a consequence, such platforms will be the basis of large drug screenings in a high-throughput manner. In addition, cell therapies with iPSC-derived differentiated adult cell types have the potential to reduce the urgent need for allogenic (that is, other human derived) organ transplants. However, before iPSC-derived tissues can be transplanted routinely, their immunogenicity needs to be investigated. Although in theory iPSCs and their derivatives should be immune-compatible to the patient from which they are derived, immunogenicity (or tolerance) of iPSC-derived cell products is still controversial (Fairchild, 2010; Fu, 2014; Tapia and Scholer, 2016). Thus, current transplantations of iPSC-derived cells are accompanied by immunosuppression to avoid possible transplant rejection. Recent reports have shown that different iPSC derivatives vary in their immunogenicity – e.g. iPSC-derived smooth muscle cells mount an immune reaction in mice while retinal epithelial cells are tolerated (Zhao et al., 2015). A future challenge will be to elucidate the reasons for these different immunological behaviours and to improve iPSC generation, cultivation, and differentiation *in vitro* in order to transplant truly immunological matched iPSC derivatives without immunosuppression and the risk of transplant rejection.

In conclusion, only a decade after Yamanaka's groundbreaking discovery, iPSC-based disease models, drug screenings, and cell therapies have found their way into clinical trials. Even though many applications are still not routinely used in the clinic, iPSC-based therapies will be a cornerstone of regenerative medicine in the future.

References

Asprer, J. S. and Lakshmipathy, U. (2015). Current methods and challenges in the comprehensive characterization of human pluripotent stem cells. *Stem Cell Reviews*, 11, pp. 357–372.

Avior, Y., Sagi, I. and Benvenisty, N. (2016). Pluripotent stem cells in disease modelling and drug discovery. *Nature Reviews Molecular Cell Biology*, 17, pp. 170–182.

Ban, H., Nishishita, N., Fusaki, N., Tabata, T., Saeki, K., Shikamura, M., Takada, N., Inoue, M., Hasegawa, M., Kawamata, S. and Nishikawa, S. (2011). Efficient generation of transgene-free human induced pluripotent stem cells (iPSCs) by temperature-sensitive Sendai virus vectors. *Proceedings of the National Academy of Sciences USA*, 108, pp. 14234–14239.

Bellin, M., Marchetto, M. C., Gage, F. H. and Mummery, C. L. (2012). Induced pluripotent stem cells: The new patient? *Nature Reviews Molecular Cell Biology*, 13, pp. 713–726.

Blau, H. M., Chiu, C. P. and Webster, C. (1983). Cytoplasmic activation of human nuclear genes in stable heterocaryons. *Cell*, 32, pp. 1171–1180.

Boroviak, T. and Nichols, J. 2014. The birth of embryonic pluripotency. *Philosophical Transactions of the Royal Society B: Biological Sciences*, 369. https://royalsociety publishing.org/doi/pdf/10.1098/rstb.2013.0541.

Briggs, R. and King, T. J. (1952). Transplantation of living nuclei from blastula cells into enucleated frogs' eggs. *Proceedings of the National Academy of Sciences USA*, 38, pp. 455–463.

Carey, B. W., Markoulaki, S., Hanna, J., Saha, K., Gao, Q., Mitalipova, M. and Jaenisch, R. (2009). Reprogramming of murine and human somatic cells using a single polycistronic vector. *Proceedings of the National Academy of Sciences USA*, 106, pp. 157–162.

Chin, M. H., Mason, M. J., Xie, W., Volinia, S., Singer, M., Peterson, C., Ambartsumyan, G., Aimiuwu, O., Richter, L., Zhang, J., Khvorostov, I., Ott, V., Grunstein, M., Lavon, N., Benvenisty, N., Croce, C. M., Clark, A. T., Baxter, T., Pyle, A. D., Teitell, M. A., Pelegrini, M., Plath, K. and Lowry, W. E. (2009). Induced pluripotent stem cells and embryonic stem cells are distinguished by gene expression signatures. *Cell Stem Cell*, 5, pp. 111–123.

Cyranoski, D. (2014a). Japanese woman is first recipient of next-generation stem cells. *Nature News*, 62, p. 3, 12 Sep. https://www.nature.com/news/japanese-woman-is-first-recipient-of-next-generation-stem-cells-1.15915.

Cyranoski, D. (2014b). Stem cells: The black box of reprogramming. *Nature*, 516, pp. 162–164.

David, L. and Polo, J. M. (2014). Phases of reprogramming. *Stem Cell Research*, 12, pp. 754–761.

Davis, R. L., Weintraub, H. and Lassar, A. B. (1987). Expression of a single transfected cDNA converts fibroblasts to myoblasts. *Cell*, 51, pp. 987–1000.

De Paepe, C., Krivega, M., Cauffman, G., Geens, M. and Van De Velde, H. (2014). Totipotency and lineage segregation in the human embryo. *Molecular Human Reproduction*, 20, pp. 599–618.

Downing, T. L., Soto, J., Morez, C., Houssin, T., Fritz, A., Yuan, F., Chu, J., Patel, S., Schaffer, D. V. and Li, S. (2013). Biophysical regulation of epigenetic state and cell reprogramming. *Nature Materials*, 12, pp. 1154–1162.

Eggan, K., Baldwin, K., Tackett, M., Osborne, J., Gogos, J., Chess, A., Axel, R. and Jaenisch, R. (2004). Mice cloned from olfactory sensory neurons. *Nature*, 428, 44–49.

Esteban, M. A., Wang, T., Qin, B., Yang, J., Qin, D., Cai, J., Li, W., Weng, Z., Chen, J., Ni, S., Chen, K., Li, Y., Liu, X., Xu, J., Zhang, S., Li, F., He, W., Labuda, K., Song, Y., Peterbauer, A., Wolbank, S., Redl, H., Zhong, M., Cai, D., Zeng, L. and Pei, D. (2010). Vitamin C enhances the generation of mouse and human induced pluripotent stem cells. *Cell Stem Cell*, 6, pp. 71–79.

Evans, M. J. and Kaufman, M. H. (1981). Establishment in culture of pluripotential cells from mouse embryos. *Nature*, 292, pp. 154–156.

Fairchild, P. J. (2010). The challenge of immunogenicity in the quest for induced pluripotency. *Nature Reviews Immunology*, 10, pp. 868–75.

Fu, X. (2014). The immunogenicity of cells derived from induced pluripotent stem cells. *Cellular and Molecular Immunology*, 11, pp. 14–16.

Fusaki, N., Ban, H., Nishiyama, A., Saeki, K. and Hasegawa, M. (2009). Efficient induction of transgene-free human pluripotent stem cells using a vector based on Sendai virus, an RNA virus that does not integrate into the host genome. *Proceedings of the Japan Academy, Series B, Physical and Biological Science*, 85, pp. 348–362.

Gaj, T., Gersbach, C. A. and Barbas, C. F., 3RD (2013). ZFN, TALEN, and CRISPR/Cas-based methods for genome engineering. *Trends in Biotechnology*, 31, pp. 397–405.

Graf, T. and Enver, T. (2009). Forcing cells to change lineages. *Nature*, 462, pp. 587–594.

Gurdon, J. B. (1962). The developmental capacity of nuclei taken from intestinal epithelium cells of feeding tadpoles. *Journal of Embryology and Experimental Morphology*, 10, pp. 622–640.

Hochedlinger, K. and Jaenisch, R. (2002). Monoclonal mice generated by nuclear transfer from mature B and T donor cells. *Nature*, 415, pp. 1035–1038.

Huangfu, D., Maehr, R., Guo, W., Eijkelenboom, A., Snitow, M., Chen, A. E. and Melton, D. A. (2008). Induction of pluripotent stem cells by defined factors is greatly improved by small-molecule compounds. *Nature Biotechnology*, 26, pp. 795–797.

Institutet, T. N. A. A. K. (2012). Press Release.

Kim, K., Zhao, R., Doi, A., Ng, K., Unternaehrer, J., Cahan, P., Huo, H., Loh, Y. H., Aryee, M. J., Lensch, M. W., Li, H., Collins, J. J., Feinberg, A. P. and Daley, G. Q. (2011). Donor cell type can influence the epigenome and differentiation potential of human induced pluripotent stem cells. *Nature Biotechnology*, 29, pp. 1117–1119.

Kulessa, H., Frampton, J. and Graf, T. (1995). GATA-1 reprograms avian myelomonocytic cell lines into eosinophils, thromboblasts, and erythroblasts. *Genes Dev*, 9, pp. 1250–1262.

Kyodo (2015). First iPSC transplant patient makes progress one year on. *Japan Times*, 2 Oct.

Lenz, M., Goetzke, R., Schenk, A., Schubert, C., Veeck, J., Hemeda, H., Koschmieder, S., Zenke, M., Schuppert, A. and Wagner, W. (2015). Epigenetic biomarker to support classification into pluripotent and non-pluripotent cells. *Scientific Reports*, 5, p. 8973.

Loi, P., Iuso, D., Czernik, M. and Ogura, A. (2016). A new, dynamic era for somatic cell nuclear transfer? *Trends in Biotechnology*, 34, pp. 791–797.

Maherali, N. and Hochedlinger, K. (2008). Guidelines and techniques for the generation of induced pluripotent stem cells. *Cell Stem Cell*, 3, pp. 595–605.

Marion, R. M., Strati, K., Li, H., Tejera, A., Schoeftner, S., Ortega, S., Serrano, M. and Blasco, M. A. (2009). Telomeres acquire embryonic stem cell characteristics in induced pluripotent stem cells. *Cell Stem Cell*, 4, pp. 141–154.

Martin, G. R. (1981). Isolation of a pluripotent cell line from early mouse embryos cultured in medium conditioned by teratocarcinoma stem cells. *Proceedings of the National Academy of Sciences USA*, 78, pp. 7634–7638.

Moran, N. (2013). Banking iPSCs. *Nature Biotechnology News*, p. 11.

Muller, F. J., Schuldt, B. M., Williams, R., Mason, D., Altun, G., Papapetrou, E. P., Danner, S., Goldmann, J. E., Herbst, A., Schmidt, N. O., Aldenhoff, J. B., Laurent, L. C. and Loring, J. F. (2011). A bioinformatic assay for pluripotency in human cells. *Nature Methods*, 8, pp. 315–317.

Mussolino, C. and Cathomen, T. (2012). TALE nucleases: Tailored genome engineering made easy. *Current Opinion in Biotechnology*, 23, pp. 644–650.

Nerlov, C. and Graf, T. (1998). PU.1 induces myeloid lineage commitment in multipotent hematopoietic progenitors. *Genes & Development*, 12, pp. 2403–2412.

Okita, K., Ichisaka, T. and Yamanaka, S. (2007). Generation of germline–competent induced pluripotent stem cells. *Nature*, 448, pp. 313–317.

Rao, M. S. and Malik, N. (2012). Assessing iPSC reprogramming methods for their suitability in translational medicine. *Journal of Cellular Biochemistry*, 113, pp. 3061–3068.

Schneuwly, S., Klemenz, R. and Gehring, W. J. (1987). Redesigning the body plan of Drosophila by ectopic expression of the homoeotic gene Antennapedia. *Nature*, 325, pp. 816–868.

Seki, T., Yuasa, S. and Fukuda, K. (2011). Derivation of induced pluripotent stem cells from human peripheral circulating T cells. *Current Protocols in Stem Cell Biology*, Chapter 4, Unit 4A 3, 18(1), pp. 4A.3.1–4A.3.9.

Seltmann, S., Lekschas, F., Muller, R., Stachelscheid, H., Bittner, M. S., Zhang, W., Kidane, L., Seriola, A., Veiga, A., Stacey, G. and Kurtz, A. (2016). hPSCreg – the human pluripotent stem cell registry. *Nucleic Acids Research*, 44, D757–D763.

Smith, Z. D., Sindhu, C. and Meissner, A. (2016). Molecular features of cellular reprogramming and development. *Nature Reviews Molecular Cell Biology*, 17, pp. 139–154.

Soldner, F., Hockemeyer, D., Beard, C., Gao, Q., Bell, G. W., Cook, E. G., Hargus, G., Blak, A., Cooper, O., Mitalipova, M., Isacson, O. and Jaenisch, R. (2009). Parkinson's disease patient-derived induced pluripotent stem cells free of viral reprogramming factors. *Cell*, 136, pp. 964–977.

Sommer, C. A., Stadtfeld, M., Murphy, G. J., Hochedlinger, K., Kotton, D. N. and Mostoslavsky, G. (2009). Induced pluripotent stem cell generation using a single lentiviral stem cell cassette. *Stem Cells*, 27, pp. 543–549.

Stadtfeld, M., Maherali, N., Breault, D. T. and Hochedlinger, K. (2008). Defining molecular cornerstones during fibroblast to iPS cell reprogramming in mouse. *Cell Stem Cell*, 2, pp. 230–240.

Tachibana, M., Amato, P., Sparman, M., Gutierrez, N. M., Tippner-Hedges, R., Ma, H., Kang, E., Fulati, A., Lee, H. S., Sritanaudomchai, H., Masterson, K., Larson, J., Eaton, D., Sadler-Fredd, K., Battaglia, D., Lee, D., Wu, D., Jensen, J., Patton, P., Gokhale, S., Stouffer, R. L., Wolf, D. and Mitalipov, S. (2013). Human embryonic stem cells derived by somatic cell nuclear transfer. *Cell*, 153, pp. 1228–1238.

Takahashi, K. and Yamanaka, S. (2006). Induction of pluripotent stem cells from mouse embryonic and adult fibroblast cultures by defined factors. *Cell*, 126, pp. 663–676.

Takahashi, K. and Yamanaka, S. (2016). A decade of transcription factor-mediated reprogramming to pluripotency. *Nature Reviews Molecular Cell Biology*, 17, pp. 183–93.

Takahashi, K., Tanabe, K., Ohnuki, M., Narita, M., Ichisaka, T., Tomoda, K. and Yamanaka, S. (2007). Induction of pluripotent stem cells from adult human fibroblasts by defined factors. *Cell*, 131, pp. 861–872.

Tapia, N. and Scholer, H. R. (2016). Molecular obstacles to clinical translation of iPSCs. *Cell Stem Cell*, 19, pp. 298–309.

Thomson, J. A., Itskovitz-Eldor, J., Shapiro, S. S., Waknitz, M. A., Swiergiel, J. J., Marshall, V. S. and Jones, J. M. (1998). Embryonic stem cell lines derived from human blastocysts. *Science*, 282, pp. 1145–1147.

Trokovic, R., Weltner, J., Manninen, T., Mikkola, M., Lundin, K., Hamalainen, R., Suomalainen, A. and Otonkoski, T. (2013). Small molecule inhibitors promote efficient generation of induced pluripotent stem cells from human skeletal myoblasts. *Stem Cells Dev*, 22, pp. 114–123.

Trounson, A. and Dewitt, N. D. (2016). Pluripotent stem cells progressing to the clinic. *Nature Reviews Molecular Cell Biology*, 17, pp. 194–200.

Tucker, B. A., Anfinson, K. R., Mullins, R. F., Stone, E. M. and Young, M. J. (2013). Use of a synthetic xeno-free culture substrate for induced pluripotent stem cell induction and retinal differentiation. *Stem Cells Translational Medicine*, 2, pp. 16–24.

Wang, H., La Russa, M. and Qi, L. S. (2016). CRISPR/Cas9 in genome editing and beyond. *Annual Review of Biochemistry*, 85, pp. 227–264.

Wilmut, I., Schnieke, A. E., Mcwhir, J., Kind, A. J. and Campbell, K. H. (1997). Viable offspring derived from fetal and adult mammalian cells. *Nature*, 385, 810–813.

Xie, H., Ye, M., Feng, R. and Graf, T. (2004). Stepwise reprogramming of B cells into macrophages. *Cell*, 117, pp. 663–676.

Yamanaka, S. and Blau, H. M. (2010). Nuclear reprogramming to a pluripotent state by three approaches. *Nature*, 465, pp. 704–712.

Yandell, K. (2016). Pluripotency Bots. *The Scientist*, 1 Jan.

Yu, J., Hu, K., Smuga-Otto, K., Tian, S., Stewart, R., Slukvin, Ii and Thomson, J. A. (2009). Human induced pluripotent stem cells free of vector and transgene sequences. *Science*, 324, pp. 797–801.

Yu, J., Vodyanik, M. A., Smuga-Otto, K., Antosiewicz-Bourget, J., Frane, J. L., Tian, S., Nie, J., Jonsdottir, G. A., Ruotti, V., Stewart, R., Slukvin, I. and Thomson, J. A. (2007). Induced pluripotent stem cell lines derived from human somatic cells. *Science*, 318, pp. 1917–1920.

Zhao, T., Zhang, Z. N., Westenskow, P. D., Todorova, D., Hu, Z., Lin, T., Rong, Z., Kim, J., He, J., Wang, M., Clegg, D. O., Yang, Y. G., Zhang, K., Friedlander, M. and Xu, Y. (2015). Humanized mice reveal differential immunogenicity of cells derived from autologous induced pluripotent stem cells. *Cell Stem Cell*, 17, pp. 353–359.

Zhu, S., Li, W., Zhou, H., Wei, W., Ambasudhan, R., Lin, T., Kim, J., Zhang, K. and Ding, S. (2010). Reprogramming of human primary somatic cells by OCT4 and chemical compounds. *Cell Stem Cell*, 7, pp. 651–655.

6 *In vivo* reprogramming

A new era in regenerative medicine

Maryam Ghasemi-Kasman

Abstract

Astrocytes' activation and resulting glial scar formation in the central nervous system (CNS) are regarded as common hallmarks of many neurodegenerative disorders. Therefore, the conversion of endogenous activated glial cells to desired cells in an injury site will be considered as the ideal strategy for the improvement of CNS repair. Recently, *in vivo* reprogramming of glial cells to neurons has opened a new route for cell replacement therapy. Here, we briefly summarize the current developments in the reprogramming of terminally differentiated cells to neuronal progenitors or neurons *in vivo*. Thereafter the advantages and challenges of *in vivo* reprogramming will be discussed.

Neuronal repair improvement in the central nervous system

Neuronal loss is a common hallmark of many neurological disorders. It has long been thought that the CNS of adult mammals does not have the ability for regeneration after injury. However, growing knowledge indicates that regeneration in the CNS might be achieved through two important strategies: (i) endogenous neural stem cells' (NSCs') activation and (ii) stem cell transplantation (Okano and Sawamoto, 2008). In adult brains NSCs exist in two discrete regions, namely the subventricular zone (SVZ) of lateral ventricles and the subgranular zone (SGZ) of the hippocampus (Doetsch et al., 1999). However, upon injury endogenous NSCs can proliferate and migrate into the damaged area and differentiate to desired cells, but the capability of the endogenous mechanism for repair is insufficient. Stem cell transplantation is considered an alternative approach for neuroregeneration improvement. It has been suggested that human embryonic stem cells (ESCs) could efficiently differentiate into neurons, and these derived neurons could be integrated into the local neuronal circuit after transplantation (Björklund et al., 2002). Despite these advantages, because of ethical concerns and immunorejection following transplantation, the application of human ESCs for clinical therapy has been hampered (De Wert and Mummery, 2003). It was long supposed that the conversion of fully differentiated cells to the embryonic state is impossible. In 1962, John Gurdon successfully transferred the cell nucleus of a frog into an enucleated oocyte which led to tadpole formation (Prasad et al., 2016). This cloning process was the first evidence that confirmed the plasticity of adult cells. For several years, many attempts were made to re-program somatic cells to the stem cell state without nuclear transference. In 2006, the discovery of induced pluripotent stem cells (iPSCs) opened a new window for regenerative medicine (Takahashi and Yamanaka, 2006).

Through this approach, by introducing the four ESC transcription factors Oct_4, Sox_2, Klf_4, and c-Myc, fibroblasts could convert to iPSCs *in vitro*. These iPSCs possess ESCs' morphology and express specific genes related to human ESCs. iPSCs' generation has opened up new horizons for cell therapy – for example, their application not only reduced immunorejection but has also been considered as a good tool for drug screening (Kim, 2014). Furthermore, iPSC technology overcame the ethical concerns related to the use of ESCs, thus becoming an attractive strategy for cell replacement therapy. Additionally, clinical applications with iPSC-derived cells have been initiated for treating several neurological disorders such as multiple sclerosis (Xie et al., 2016), Parkinson's Disease (PD) (Nishimura and Taka-hashi, 2013), and spinal cord injury (Nakamura and Okano, 2013). Despite the advantages of iPSCs in regenerative therapies, their ability for teratoma formation and epigenetic variations of induced cells has hampered their application in the clinic (Lee et al., 2013). To overcome the limitations of iPSCs for cell therapy, direct reprogramming has been introduced as an al-ternative strategy in the regenerative medicine field. A wide range of somatic cells such as fibroblasts and astrocytes can successfully convert into neurons using specific transcription factors, microRNAs, and small molecules in a culture condition (Gopalakrishnan et al., 2017). More recently, the *in vivo* reprogramming of somatic cells into desired cells has been regarded as an ideal and promising strategy for cell therapy. A variety of somatic cells has been converted to target cells using the *in vivo* application of transcription factors or microRNAs (Smith and Zhang, 2015). In this chapter, we will highlight the recent progress of *in vivo* reprogramming in the CNS and we will show the major challenges that induced neurons present for clinical applications.

Direct lineage reprogramming

Lineage reprogramming is defined as the conversion of somatic cells to another type of cell without passing from the pluripotent stage (Gopalakr-ishnan et al., 2017). In recent years, extensive studies have shown that a variety of differentiated cell types can be converted into other terminally differentiated cells. As a result, the lineage reprogramming approach has become one of the most promising strategies for the generation of func-tional cell types. For the first time, the possibility of the direct conver-sion of somatic cells was shown by Davis and his colleagues in 1987 (Davis et al., 1987). Their results indicated that fibroblasts can be converted to my-oblasts by the overexpression of specific muscle transcription factors such as *Myod*. Subsequently, several studies indicated that fibroblasts can be directly reprogrammed into other cell types such as cardiomyocytes (Ieda et al., 2010), hepatocytes (Huang et al., 2014), and oligodendrocytes (Najm et al., 2013).

Neuronal loss is a common hallmark of many neurodegenerative diseases which can lead to functional impairments (Guo et al., 2014). Recently, the reprogramming of somatic cells to neurons has been introduced as a novel strategy in the enhancement of endogenous repair. The earliest report, by Vierbuchen et al., demonstrated that both embryonic and postnatal fibroblasts can be efficiently converted to neurons using the overexpression of neural lineage-specific transcription factors such as *Ascl1*, *Brn2*, and *Myt1* (*ABM*). These induced neurons expressed a specific neuronal marker and had the ability for action potential firing (Vierbuchen et al., 2010). Further study by this group also showed that human fibroblasts converted to functional neurons by the introduction of *ABM* and *NeuroD1* transcription factors into a culture medium (Pang et al., 2011). Interestingly, the *Ascl1*-induced neurons were mostly excitatory – revealing the generation of certain subtypes of neuron through this approach. In addition to the ectopic expression of transcription factors, neuronal induction can be achieved using certain microRNAs and small molecules (Chen et al., 2016). However, though the overexpression of neuronal-specific microRNAs such as miR-9/9* and miR-124 could convert human fibroblasts to neurons, the efficiency of this process was low and the overexpression of transcription factors was required to obtain functional neurons (Yoo et al., 2011). Delivery of transcription factors and microRNAs through viral injection has been noticed as the main problem for using this approach in patients (Li and Chen, 2016). An alternative method is the direct conversion of cells by the specific combinations of small molecules. Interestingly, Li et al. could convert 90% of fibroblasts to neurons using a combination of four small molecules (Li et al., 2015). Additionally, astrocytes have also been converted to neurons by a combination of small molecules (Zhang et al., 2015). Furthermore, induced neurons can be derived from the human fibroblasts of Alzheimer's disease (AD) patients using small molecules (Hu et al., 2015). Despite the beneficial effects of chemical reprogramming, there is no possibility for the application of small molecules *in vivo* – and the induced cells have to be transplanted into an adult brain which still faces the hurdle of immunorejection (Li and Chen, 2016).

Besides fibroblasts, a variety of somatic cells including hepatocytes (Marro et al., 2011), adipocytes (Yang et al., 2013), and glial cells such as astrocytes have been converted to neurons in a culture condition (Corti et al., 2012). Subsequently, different neuronal subtypes such as dopaminergic neurons (Di Val Cervo et al., 2017), motor neurons (Son et al., 2011), and retinal ganglion cells (Chen et al., 2010) have been obtained using a direct reprogramming approach. Compared to iPSCs, these induced neurons are considered as ideal candidates for the improvement of CNS repair. The induced neurons efficiently generate repetitive action potential and can form functional synapses with endogenous neurons. Additionally, the process of direct reprogramming is fast and induced cells reduce the tumourigenesis which was associated with iPSCs' transplantation. Despite the advantage of direct reprogramming in a culture condition, successful integration of transplanted induced cells into local circuits is still considered to be the main limitation of this approach. Furthermore, most of the engrafted cells are immune rejected and cannot effectively reduce functional impairments following CNS injuries (Li and Chen, 2016). To overcome these limitations, recent studies have focussed on *in vivo* reprogramming of endogenous cells to neurons.

History of *in vivo* reprogramming

Recently, *in vivo* reprogramming has generated considerable interest in the regenerative medicine field. In this approach, resident cells can be directly converted to target cells by the ectopic expression of specific transcription factors or microRNAs. An earlier study by Zhou and his colleagues showed the possibility of *in vivo* reprogramming in the pancreas. They could successfully convert pancreatic exocrine cells into insulin-secreting β cells using three transcription factors. These induced β cells not only were similar to endogenous cells morphologically, but they also could partially reduce the blood glucose levels in diabetic mice (Zhou et al., 2008). A subsequent study in mouse embryos showed that mesoderm tissue can be converted to cardiac tissue using specific cardiac transcription factors (Takeuchi and Bruneau, 2009). Further studies demonstrated that induced-cardiac cells can be generated from mouse fibroblasts by introducing transcription factors *in vivo* (Qian et al., 2012; Song et al., 2012). These cardiac-like cells could successfully fire action potential, and they reduced the damage size of myocardial infarction in animal models. In addition, several *in vivo* reprogramming studies into the heart have been carried out by differing the combinations of transcription factors (Ieda et al., 2010; Sadahiro et al., 2015). Recent studies also indicate that hepatic myofibroblasts can be reprogrammed to hepatocytes *in vivo* and that these induced cells effectively attenuate liver fibrosis (Rezvani et al., 2016; Song et al., 2016).

In vivo reprogramming in the central nervous system

The conversion of resident glial cells to target cells in damaged areas by the *in vivo* lineage reprogramming approach has been recently regarded as one of the major goals of regenerative medicine. Glial cells, including astrocytes and microglia, are broadly distributed throughout the CNS (Li and Chen, 2016). In response to injury, these cells will activate and form a barrier so to restrict the spread of damage. However, though glial cell activation initially exerts this beneficial effect in restricting a damaged area, ultimately the secretion of neuroinhibitory factors by activated glial cells leads to axonal loss (Niu et al., 2013). In recent years, a growing number of studies has shown that astrocytes can be reprogrammed to neurons *in vivo*. Torper et al. transfected the human astrocytes and fibroblasts with a specific cocktail of transcription factors, and then engineered cells were transplanted into the striatum. Their results demonstrated that engrafted cells can be converted to functional neurons *in vivo* (Torper et al., 2013). Another study by the Zhang group indicated that under ectopic expression of the *Sox2* transcription factor, resident astrocytes in both the spinal cord and the striatum of adult brains were converted into neuroblasts. Interestingly, their results suggested that the presence of valproic acid is essential for the differentiation of neuroblasts to mature neurons (Niu et al., 2013; Su et al., 2014). In parallel, Guo and colleagues showed that reactive astrocytes and NG2 cells in the cortex can be reprogrammed to neurons by neurogenic transcription factors such as *NeuroD1* in an AD animal model (Guo et al., 2014). More recently, adult astrocytes have been converted to dopaminergic neurons by the *in vivo* administration of *NeuroD1*, *ASCL1*, and *LMX1A* transcription factors and microRNA 218 into the striatum of mice. These specific subtypes of induced neurons were excitatory and could reverse some behavioural impairment which occurs in PD (Di Val Cervo et al., 2017). In addition to transcription factors, reprogramming can be mediated through specific microRNAs. The miR-302/367 cluster has a high expression in ESCs and its overexpression enhances the efficiency of pluripotent reprogramming by a hundred times in comparison to the use of Yamanaka factors (Anokye-Danso et al., 2011). Our previous work indicated that when miR-302/367 lentiviral particles are injected into the striatum in a stab injury model, astrocytes as a major population of transfected cells convert into neuroblasts and neurons in the presence of valproic acid as an epigenetic modifier. Interestingly, our *in vitro* study on cultured human astrocytes also demonstrated that induced neurons were mainly excitatory and that they had an ability for repetitive action potential firing (Ghasemi-Kasman et al., 2015).

In vivo reprogramming advantages

In contrast to other animal kingdoms, the capacity of the mammalian brain and spinal cord for endogenous repair is very low (Li and Chen, 2016). Several approaches have been made to improve the self-regeneration of the CNS. Among them, *in vivo* reprogramming has emerged as a novel route for the improvement of endogenous repair in the CNS in recent years (Xu et al., 2015). In this approach, by *in vivo* administration of the reprogramming factors, internal glial cells can be reprogrammed to neurons. Unlike external stem cells, the transplantation which was frequently associated with immunorejection and tumour formation, *in vivo* reprogramming generates neurons using internal glial cells in a damaged area. For CNS repair improvement, the efficiency of neuronal generation should be high. Recent evidence has suggested that the efficiency of *in vivo* reprogramming in neuronal induction is more than 90% when the NeuroD1 transcription factor has been administered into the brain. A generation of neurons using activated glial cells, directly at the injury site, is another advantage of *in vivo* lineage reprogramming (Li and Chen, 2016). In addition, the release of several local factors from the environment (niche) exerts a significant effect on a cell's identity during the *in vivo* reprogramming process. For example, our previous work showed that when miR-302/367 lentiviral particles were injected into the striatum, the astrocytes' transduced cells were mostly converted to neurons (Ghasemi-Kasman et al., 2015); while the injection of miR-302/367 into the *corpus callosum* generated oligodendrocytes' precursor cells which led to myelin repair enhancement in the local demyelination-induced model (Ghasemi-Kasman et al., 2017).

The targeting of proliferative glial cells is another important advantage of *in vivo* conversion. However, it has been well documented that a wide range of somatic cells can be converted to neurons, but cells with self-renewal potency are considered to be the ideal source for therapeutic approaches. Activation of glial cells and resulting glial scar formation is associated with many neurological disorders including AD, PD, and CNS injuries such as spinal cord injury. Interestingly, the administration of reprogramming factors could effectively convert activated glial cells to functional neurons in animal models of AD (Guo et al., 2014), PD (Di Val Cervo et al., 2017), and spinal cord injury (Su et al., 2014). The successful *in vivo* conversion of reactive glial cells to neurons may have a significant impact on the treatment of neurodegenerative diseases in the future.

In vivo reprogramming challenges

Despite the rapid progress of *in vivo* lineage reprogramming in recent years, the procedure still faces certain hurdles which should be solved before its application in clinical therapy. The first challenge is whether newly *in vivo* reprogrammed neurons can survive in a damaged area (Li and Chen, 2016). Since induced neurons are surrounded with reactive glial cells, the release of neuroinhibitory factors by activated glial cells may eventually lead to neuronal damage. Several attempts such as neurotropic factor administration (Grande et al., 2013) or the co-expression of transcription factors with anti-apoptotic factors (Gascón et al., 2016) successfully enhanced the survival of newly reprogrammed neurons. Another important challenge is the generation of a desired subtype of neurons for treating the specific types of neurodegenerative disorders. Previous evidence indicated that glutamatergic and GABAergic neurons can be derived from glial cells (Guo et al., 2014). More recently, the generation of dopaminergic neurons has been obtained in the PD model (Di Val Cervo et al., 2017). Despite these advantages, the successful induction of other subtypes of neuronal cells such as serotonergic, norepinephrinergic, and cholinergic neurons has not still obtained *in vivo* (Li and Chen, 2016). Rebuilding the neural circuits by newly reprogrammed neurons is considered to be the most important challenge for the clinical application of *in vivo* reprogramming. Of considerable importance is the question as to whether newly converted neurons can send their projections to the right area: that which was previously innervated by the lost neurons. A crucial hurdle is whether newly induced neurons can restore the CNS function which was impaired following injury. Lately, the field of *in vivo* reprogramming has progressed not only in the discovery of new reprogramming factors but also in the novel strategies that have been developed to improve the functional maturation of induced neurons. The maturation process of induced neurons that leads to functional heterogeneity is very slow and inefficient (Xu et al., 2015). Functional heterogeneity may cause remarkable side effects. For example, electrical heterogeneity in reprogrammed cardiomyocytes may result in arrhythmias (Xu et al., 2012). Additionally, the generation of sufficient desired cell types during *in vivo* reprogramming and the ability of induced cells for successful synaptic formation with endogenous neurons also play important roles in the functional improvement in the CNS (Li and Chen, 2016).

Introducing the reprogramming factors using viral particles is another major obstacle of *in vivo* conversion which has hampered its clinical application (Xu et al., 2015; Gopalakrishnan et al., 2017). The use of viral particles may lead to insertional mutagenesis, immunogenicity, and the uncontrolled expression of reprogramming genes (Gopalakrishnan et al., 2017). Numerous strategies such as transient transfection, non-integrating viral vectors, and protein transduction have evolved to overcome the mentioned limitations of *in vivo* reprogramming (Hu, 2014). Chemical reprogramming using small molecules has been introduced as a safe and alternative strategy for direct conversion (Xu et al., 2015; Gopalakrishnan et al., 2017). Although the chemical reprogramming of cells to neurons has been achieved *in vitro*, the toxicity and side effects of these chemical compounds have limited the *in vivo* application of small molecules. Other major concerns about small molecules include the issue of whether they can pass from the blood brain barrier and act specifically on activated glial cells or not. An additional challenge is whether a single small molecule can reprogram somatic cells to neurons. Previous evidence has suggested that small molecules could reprogram somatic cells by a cocktail of several chemical substances. To minimize the cell toxicity effect of small molecules, reducing the number of small molecules is necessary for its *in vivo* administration (Li and Chen, 2016).

Conclusion

In recent years, the *in vivo* reprogramming of fully-differentiated cells has been introduced as an effective strategy in regenerative medicine. With this approach, internal reactive glial cells in an injury site can be effectively converted to target cells. Compared to conventional stem cell therapy, *in vivo* reprogramming greatly reduces tumour formation and immunorejection. Despite several prospects of *in vivo* reprogramming, significant challenges lie ahead and many issues should be addressed before the application of an *in vivo* reprogramming strategy in clinical therapies.

References

Anokye-Danso, F., Trivedi, C. M., Juhr, D., Gupta, M., Cui, Z., Tian, Y., Zhang, Y., Yang, W., Gruber, P. J. and Epstein, J. A. (2011). Highly efficient miRNA-mediated reprogramming of mouse and human somatic cells to pluripotency. *Cell Stem Cell*, 8, pp. 376–388.

Björklund, L. M., Sánchez-Pernaute, R., Chung, S., Andersson, T., Chen, I. Y. C., Mcnaught, K. S. P., Brownell, A.-L., Jenkins, B. G., Wahlestedt, C. and Kim, K.-S. (2002). Embryonic stem cells develop into functional dopaminergic neurons after transplantation in a Parkinson rat model. *Proceedings of the National Academy of Sciences USA*, 99, pp. 2344–2349.

Chen, M., Chen, Q., Sun, X., Shen, W., Liu, B., Zhong, X., Leng, Y., Li, C., Zhang, W. and Chai, F. (2010). Generation of retinal ganglion-like cells from reprogrammed mouse fibroblasts. *Investigative Ophthalmology & Visual Science*, 51, pp. 5970–5978.

Chen, Y., Pu, J. and Zhang, B. (2016). Progress and challenges of cell replacement therapy for neurodegenerative diseases based on direct neural reprogramming. *Human Gene Therapy*, 27, pp. 962–970.

Corti, S., Nizzardo, M., Simone, C., Falcone, M., Donadoni, C., Salani, S., Rizzo, F., Nardini, M., Riboldi, G. and Magri, F. (2012). Direct reprogramming of human astrocytes into neural stem cells and neurons. *Experimental Cell Research*, 318, pp. 1528–1541.

Davis, R. L., Weintraub, H. and Lassar, A. B. (1987). Expression of a single transfected cDNA converts fibroblasts to myoblasts. *Cell*, 51, pp. 987–1000.

De Wert, G. and Mummery, C. (2003). Human embryonic stem cells: Research, ethics and policy. *Human Reproduction*, 18, pp. 672–682.

Di Val Cervo, P. R., Romanov, R. A., Spigolon, G., Masini, D., Martín-Montañez, E., Toledo, E. M., La Manno, G., Feyder, M., Pifl, C. and Ng, Y.-H. (2017). Induction of functional dopamine neurons from human astrocytes in vitro and mouse astrocytes in a Parkinson's disease model. *Nature Biotechnology*, 35, pp. 444–452.

Doetsch, F., Caille, I., Lim, D. A., Garcia-Verdugo, J. M. and Alvarez-Buylla, A. (1999). Subventricular zone astrocytes are neural stem cells in the adult mammalian brain. *Cell*, 97, pp. 703–716.

Gascón, S., Murenu, E., Masserdotti, G., Ortega, F., Russo, G. L., Petrik, D., Deshpande, A., Heinrich, C., Karow, M. and Robertson, S. P. (2016). Identification and successful negotiation of a metabolic checkpoint in direct neuronal reprogramming. *Cell Stem Cell*, 18, pp. 396–409.

Ghasemi-Kasman, M., Hajikaram, M., Baharvand, H. and Javan, M. (2015). MicroRNA-mediated in vitro and in vivo direct conversion of astrocytes to neuroblasts. *PloS One*, 10, e0127878.

Ghasemi-Kasman, M., Zare, L., Baharvand, H. and Javan, M. (2018). In vivo conversion of astrocytes to myelinating cells by miR-302/367 and valproate to enhance myelin repair. *Journal of Tissue Engineering and Regenerative Medicine*. 12(1), pp. e462–e472.

Gopalakrishnan, S., Hor, P. and Ichida, J. K. (2017). New approaches for direct conversion of patient fibroblasts into neural cells. *Brain Research*, 1656, pp. 2–13.

Grande, A., Sumiyoshi, K., López-Juárez, A., Howard, J., Sakthivel, B., Aronow, B., Campbell, K. and Nakafuku, M. (2013). Environmental impact on direct neuronal reprogramming in vivo in the adult brain. *Nature Communications*, 4, pp. 1–26.

Guo, Z., Zhang, L., Wu, Z., Chen, Y., Wang, F. and Chen, G. (2014). In vivo direct reprogramming of reactive glial cells into functional neurons after brain injury and in an Alzheimer's disease model. *Cell Stem Cell*, 14, pp. 188–202.

Hu, K. (2014). All roads lead to induced pluripotent stem cells: the technologies of iPSC generation. *Stem Cells and Development*, 23, pp. 1285–1300.

Hu, W., Qiu, B., Guan, W., Wang, Q., Wang, M., Li, W., Gao, L., Shen, L., Huang, Y. and Xie, G. (2015). Direct conversion of normal and Alzheimer's disease human fibroblasts into neuronal cells by small molecules. *Cell Stem Cell*, 17, pp. 204–212.

Huang, P., Zhang, L., Gao, Y., He, Z., Yao, D., Wu, Z., Cen, J., Chen, X., Liu, C. and Hu, Y. (2014). Direct reprogramming of human fibroblasts to functional and expandable hepatocytes. *Cell Stem Cell*, 14, pp. 370–384.

Ieda, M., Fu, J.-D., Delgado-Olguin, P., Vedantham, V., Hayashi, Y., Bruneau, B. G. and Srivastava, D. (2010). Direct reprogramming of fibroblasts into functional cardiomyocytes by defined factors. *Cell*, 142, pp. 375–386.

Kim, C. (2014). Disease modeling and cell based therapy with iPSC: Future therapeutic option with fast and safe application. *Blood Research*, 49, pp. 7–14.

Lee, A. S., Tang, C., Rao, M. S., Weissman, I. L. and Wu, J. C. (2013). Tumorigenicity as a clinical hurdle for pluripotent stem cell therapies. *Nature Medicine*, 19, pp. 998–1004.

Li, H. and Chen, G. (2016). In vivo reprogramming for CNS repair: Regenerating neurons from endogenous glial cells. *Neuron*, 91, pp. 728–738.

Li, X., Zuo, X., Jing, J., Ma, Y., Wang, J., Liu, D., Zhu, J., Du, X., Xiong, L. and Du, Y. (2015). Small-molecule-driven direct reprogramming of mouse fibroblasts into functional neurons. *Cell Stem Cell*, 17, pp. 195–203.

Marro, S., Pang, Z. P., Yang, N., Tsai, M.-C., Qu, K., Chang, H. Y., Sudhof, T. C. and Wernig, M. (2011). Direct lineage conversion of terminally differentiated hepatocytes to functional neurons. *Cell Stem Cell*, 9, pp. 374–382.

Najm, F. J., Lager, A. M., Zaremba, A., Wyatt, K., Caprariello, A. V., Factor, D. C., Karl, R. T., Maeda, T., Miller, R. H. and Tesar, P. J. (2013). Transcription factor-mediated reprogramming of fibroblasts to expandable, myelinogenic oligodendrocyte progenitor cells. *Nature Biotechnology*, 31, pp. 426–433.

Nakamura, M. and Okano, H. (2013). Cell transplantation therapies for spinal cord injury focusing on induced pluripotent stem cells. *Cell Research*, 23, pp. 70–80.

Nishimura, K. and Takahashi, J. (2013). Therapeutic application of stem cell technology toward the treatment of Parkinson's disease. *Biological and Pharmaceutical Bulletin*, 36, pp. 171–175.

Niu, W., Zang, T., Zou, Y., Fang, S., Smith, D. K., Bachoo, R. and Zhang, C.-L. (2013). In vivo reprogramming of astrocytes to neuroblasts in the adult brain. *Nature Cell Biology*, 15, pp. 1164–1175.

Okano, H. and Sawamoto, K. (2008). Neural stem cells: Involvement in adult neurogenesis and CNS repair. *Philosophical Transactions of the Royal Society of London B: Biological Sciences*, 363, pp. 2111–2122.

Pang, Z. P., Yang, N., Vierbuchen, T., Ostermeier, A., Fuentes, D. R., Yang, T. Q., Citri, A., Sebastiano, V., Marro, S. and Sudhof, T. C. (2011). Induction of human neuronal cells by defined transcription factors. *Nature*, 476, pp. 220–223.

Prasad, A., Manivannan, J., Loong, D. T., Chua, S. M., Gharibani, P. M. and All, A. H. (2016). A review of induced pluripotent stem cell, direct conversion by trans-differentiation, direct reprogramming and oligodendrocyte differentiation. *Regenerative Medicine*, 11, pp. 181–191.

Qian, L., Huang, Y., Spencer, C. I., Foley, A., Vedantham, V., Liu, L., Conway, S. J., Fu, J.-D. and Srivastava, D. (2012). In vivo reprogramming of murine cardiac fibroblasts into induced cardiomyocytes. *Nature*, 485, pp. 593–598.

Rezvani, M., Español-Suñer, R., Malato, Y., Dumont, L., Grimm, A. A., Kienle, E., Bindman, J. G., Wiedtke, E., Hsu, B. Y. and Naqvi, S. J. (2016). In vivo hepatic reprogramming of myofibroblasts with AAV vectors as a therapeutic strategy for liver fibrosis. *Cell Stem Cell*, 18, pp. 809–816.

Sadahiro, T., Yamanaka, S. and Ieda, M. (2015). Direct cardiac reprogramming. *Circulation Research*, 116, pp. 1378–1391.

Smith, D. K. and Zhang, C.-L. (2015). Regeneration through reprogramming adult cell identity in vivo. *The American Journal of Pathology*, 185, pp. 2619–2628.

Son, E. Y., Ichida, J. K., Wainger, B. J., Toma, J. S., Rafuse, V. F., Woolf, C. J. and Eggan, K. (2011). Conversion of mouse and human fibroblasts into functional spinal motor neurons. *Cell Stem Cell*, 9, pp. 205–218.

Song, G., Pacher, M., Balakrishnan, A., Yuan, Q., Tsay, H.-C., Yang, D., Reetz, J., Brandes, S., Dai, Z. and Putzer, B. M. (2016). Direct reprogramming of hepatic myofibroblasts into hepatocytes in vivo attenuates liver fibrosis. *Cell Stem Cell*, 18, pp. 797–808.

Song, K., Nam, Y.-J., Luo, X., Qi, X., Tan, W., Huang, G. N., Acharya, A., Smith, C. L., Tallquist, M. D. and Neilson, E. G. (2012). Heart repair by reprogramming non-myocytes with cardiac transcription factors. *Nature*, 485, pp. 599–604.

Su, Z., Niu, W., Liu, M.-L., Zou, Y. and Zhang, C.-L. (2014). In vivo conversion of astrocytes to neurons in the injured adult spinal cord. *Nature Communications*, 5, pp. 1–15.

Takahashi, K. and Yamanaka, S. (2006). Induction of pluripotent stem cells from mouse embryonic and adult fibroblast cultures by defined factors. *Cell*, 126, pp. 663–676.

Takeuchi, J. K. and Bruneau, B. G. (2009). Directed transdifferentiation of mouse mesoderm to heart tissue by defined factors. *Nature*, 459, pp. 708–711.

Torper, O., Pfisterer, U., Wolf, D. A., Pereira, M., Lau, S., Jakobsson, J., Björklund, A., Grealish, S. and Parmar, M. (2013). Generation of induced neurons via direct conversion in vivo. *Proceedings of the National Academy of Sciences USA*, 110, pp. 7038–7043.

Vierbuchen, T., Ostermeier, A., Pang, Z. P., Kokubu, Y., Sudhof, T. C. and Wernig, M. (2010). Direct conversion of fibroblasts to functional neurons by defined factors. *Nature*, 463, pp. 1035–1041.

Xie, C., Liu, Y.-Q., Guan, Y.-T. and Zhang, G.-X. (2016). Induced stem cells as a novel multiple sclerosis therapy. *Current Stem Cell Research & Therapy*, 11, pp. 313–320.

Xu, H., Yi, B. A. and Chien, K. R. (2012). In vivo reprogramming for heart disease. *Cell Research*, 22, pp. 1521–1523.

Xu, J., Du, Y. and Deng, H. (2015). Direct lineage reprogramming: Strategies, mechanisms, and applications. *Cell Stem Cell*, 16, pp. 119–134.

Yang, Y., Jiao, J., Gao, R., Yao, H., Sun, X.-F. and Gao, S. (2013). Direct conversion of adipocyte progenitors into functional neurons. *Cellular Reprogramming (Formerly "Cloning and Stem Cells")*, 15, pp. 484–489.

Yoo, A. S., Sun, A. X., Li, L., Shcheglovitov, A., Portmann, T., Li, Y., Lee-Messer, C., Dolmetsch, R. E., Tsien, R. W. and Crabtree, G. R. (2011). MicroRNA-mediated conversion of human fibroblasts to neurons. *Nature*, 476, pp. 228–231.

Zhang, L., Yin, J.-C., Yeh, H., Ma, N.-X., Lee, G., Chen, X. A., Wang, Y., Lin, L., Chen, L. and Jin, P. (2015). Small molecules efficiently reprogram human astroglial cells into functional neurons. *Cell Stem Cell*, 17, pp. 735–747.

Zhou, Q., Brown, J., Kanarek, A., Rajagopal, J. and Melton, D. A. (2008). In vivo reprogramming of adult pancreatic exocrine cells to beta-cells. *Nature*, 455, pp. 627–632.

7 Modelling human neurodegeneration using induced pluripotent stem cells

Iryna Prots, Beate Winner, and Jürgen Winkler

Abstract

Neurodegeneration is a process of neuronal damage and death. A progressive loss of neurons is a common pathological feature of many neurodegenerative diseases, including motor neuron diseases (MND) and Parkinson's disease (PD). Neurodegenerative diseases are incurable and the only therapies possible are to this date symptomatic. Discovering the mechanisms that either lead to the disease occurrence or that influence the disease progression (together called 'pathomechanisms') would open up possibilities for developing new therapeutic approaches. Cellular material for research in the field of human neurodegeneration is very limited and mostly available as a post-mortem tissue derivate. Therefore, stem cells, as a source of different types of somatic cells, are a very promising tool for developing suitable human models to investigate neurodegeneration. With the discovery of induced pluripotent stem cells (iPSC), researchers now have access to otherwise difficult-to-obtain patient-specific cellular material to investigate the mechanisms that cause neuronal loss and that lead to the development of a neurodegenerative disease. In other words, iPSC-based technology allows the retrieving of disease-specific neurons and thereby allows the observation of the disease development in a dish. In this article, we first describe the symptomatic and current management of two big groups of neurodegenerative diseases: MND and PD. We next explain the advantages of using iPSC technology for the investigation of human neurodegeneration. Then, we highlight a number of pathomechanisms underlying neuronal degeneration in MND and PD, which have been uncovered in iPSC-based models. We finally show that some disease-related phenotypes could already be rescued on a cellular level in the laboratory and discuss the iPSC-based research as a promising tool, fuelling hope that new successful therapies for neurodegenerative disorders can be developed.

Human neurodegenerative diseases

Neurodegeneration

Neurodegenerative diseases, including motor neuron disease (MND) and Parkinson's disease (PD, affect millions of people worldwide in ways distressingly disruptive to their victims and their kin. Neurodegenerative diseases are defined as conditions that primarily affect neurons.

Neurons are the building blocks of the nervous system. Neurons can undergo damage or cell death. A process of structural or functional damage of neurons up to neuronal death is termed *neurodegeneration*. Progressive loss of neurons in the human brain and spinal cord is a common pathological feature of a number of neurodegenerative diseases (Sheikh et al., 2013). As a result, problems with movement or mental functioning occur as symptoms, depending on the type of degenerating neuronal cell. Generally, neurodegenerative diseases can occur sporadically or can be inherited, they are incurable, and they have unknown triggers and largely unclear mechanisms (ibid.). MND and PD are caused by the loss of neurons in different regions of the central nervous system.

Motor neuron diseases

MND is a heterogeneous group of neurological disorders that causes rapidly progressive muscle weakness, affecting a person's ability to walk, speak, swallow, and breathe (Worms, 2001). MND typically affects people in their mid-fifties. MND primarily attacks motor neurons, the neurons that control essential voluntary muscles of the body. Motor neurons consist of two types: the corticospinal motor neurons (or upper motor neurons) and the alpha motor neurons (or lower motor neurons). Motor neurons are responsible for the transmission of the movement commands from the motor cortex in the brain to the spinal cord (upper motor neurons), and from the spinal cord to the muscle fibres (lower motor neurons). The activation of muscle fibres by lower motor neurons leads to the contraction of the muscle, resulting in the voluntary movement of the body.

A subgroup of MND that primarily affects the upper motor neurons is *hereditary spastic paraplegia* (HSP) (Blackstone et al., 2011). The prevalence of HSP is between 3 and 12 cases per 100,000 population in Europe, and the age of onset can vary widely from early childhood through to late in life – making HSP a significant source of chronic neurodisability throughout society. HSP is an inherited disease, and a clinically and genetically heterogeneous group of monogenic MND. Mutations in seventy-four different gene loci are known to cause HSP and are called spastic paraplegia genes (SPG) 1–71 (Fink, 2013; Novarino et al., 2014). Clinically, HSP is characterized by a progressive paresis and spasticity of the lower limbs (Fink, 2013). Spasticity is thought to be the result of the impaired ability of damaged or lost upper motor neurons to regulate muscle excitability and the inhibition of muscle stretch reflexes (Kandel and Squire, 2000). The progressive nature of HSP makes patients dependent on canes or Zimmer frames, and more severe cases require a wheelchair later in life.

HSP can be classified into two groups: pure and complicated, based on the absence (pure) or presence (complicated) of additional clinical features besides spastic paraparesis, such as ataxia, extrapyramidal signs, peripheral neuropathy, epilepsy, deafness, optic atrophy, pigmentary retinopathy, cognitive impairment, dementia, and severe amyotrophy (reviewed in [Blackstone, 2012; Fink, 2013]). Unfortunately, there are still no medications or therapies which cure or at least halt disease progression in HSP. The existing therapies are only symptomatic and allow the reduction of symptoms of spasticity by the use of muscle relaxants or botulinum toxin (also known as Botox) to help reduce the spasticity locally (reviewed in [Fink, 2013; Soderblom and Blackstone, 2006]). Other symptoms are treated symptomatically – exercise and physical therapies are important. Therapeutic progress relies on the knowledge of the disease mechanisms. Therefore, studying HSP provides an important means, firstly, to understand the specific molecular mechanisms underlying axonal maintenance and degeneration in motor neurons and, secondly, to develop new effective therapies. It is hoped that the stem cell technology detailed in the section below on 'iPSC as a modelling tool for human neurodegeneration' will enable detailed investigations of disease mechanisms and thus treatment of HSP as well as PD.

Parkinson's disease (PD)

PD is the most common neurodegenerative movement disorder that results in the inability of a person to control movement normally (Kalia and Lang, 2015). PD may start with a tremor in a hand and develops chronically and progressively, meaning that symptoms continue and worsen over time. The prevalence of PD is 84 cases per 100 000 individuals, and it predominantly affects individuals over the age of sixty-five (WHO, 2006). Thus, PD is an enormous economic health challenge: the total annual cost of PD is more than double that of the control population with 50% of costs arising due to the productivity loss of PD patients (Huse et al., 2005). The cause of PD is unknown. In some rare cases, PD might be inherited due to mutations in several genetic loci called PARK genes (Klein and Westenberger, 2012). Genetic forms of PD have an early onset and account for only 3%–5% of PD cases.

Several important neuronal circuits between different regions of the nervous system are disrupted in PD, causing a variety of symptoms that can be divided into motor- and non-motor symptoms. The symptoms of PD vary from person to person. The four most common motor symptoms of the disease are bradykinesia or slowness of movement; rigidity or stiffness of the limbs and torso; tremor of the hands, arms, legs, jaw, and face; and postural instability or impaired balance and coordination (Kalia and Lang, 2015). PD motor symptoms are mostly attributed to a loss of neurons in an area of the mid-brain called the 'substantia nigra'. Neurons in the substantia nigra produce dopamine, a chemical that sends messages to another part of the brain (striatum) that controls movement and coordination. As PD progresses, the nigro-striatal connections become disrupted and the amount of dopamine produced in the brain decreases, leaving a person unable to control movement normally. In addition, non-motor symptoms – such as dysregulated smell and sleep, depression, mild cognitive impairment, or constipation – affect many PD patients and are increasingly recognized by doctors as important to treat. Non-motor symptoms result from the loss of neurons in other areas of the nervous system – including the brain stem, the olfactory bulb, and the enteric nervous system (neurons in the gastrointestinal tract) – and are experienced by PD patients long before any motor signs of the disease appear. Non-motor symptoms occurring prior to motor symptoms belong to the very early 'pre-diagnostic' phase of PD (Kalia and Lang, 2015).

To date, PD can be diagnosed based only on the presence of motor symptoms. But at this stage, around 80% of the dopaminergic neurons in the substantia nigra are already lost. This makes the treatment of the disease very challenging. The main therapy is still based on findings of the early 1960s: the use of dopaminergic agents to restore the balance of neurotransmission (Connolly and Lang, 2014). When the drug therapy is inefficient, and in severe cases, surgery is performed by the use of microelectrodes for deep brain stimulation to reduce motor symptoms (Hickey and Stacy, 2016). Diet and certain forms of rehabilitation have shown some effectiveness in improving symptoms as well. Thus, the clinical situation for the treatment of PD today is symptomatic and it is only able to slow the disease progression in a limited way, if at all.

Current belief is that better understanding the sequence of events that leads to the loss of dopaminergic neurons could provide us with the necessary knowledge to develop new successful treatments of PD. At present, several relevant pathological characteristics are known. The main neuropathological hallmark of PD is a formation of aggregates (clumps) of a protein alpha-synuclein, which are also called Lewy Bodies or Lewy Neurites, depending on their location in the neuronal cell: cell body or axon, respectively (Spillantini et al., 1997). The appearance of alpha-synuclein aggregates in neurons in different areas of the brain at different disease stages correlates with respective symptoms and with the time of their manifestation. Additionally, the loss of axons and dendrites very early during the course of the disease results in reduced or lost neuronal connectivity between different brain regions (Kalia and Lang, 2015). Finally, inflammatory processes due to the activation of brain immune cells, such as microglia, accompany and enhance neuronal degeneration thereby modulating disease progression. Each of these processes leads to neuronal death, but how they collude to result in the above-described PD symptoms needs to be understood.

Common challenges for the successful treatment of human neurodegenerative diseases

Three different strategies might be utilized to treat neurodegenerative diseases: symptomatic treatment, disease-modifying treatment, and the replacement of lost functions. For the majority of neurodegenerative diseases, as stated above, only symptomatic treatment exists and a common problem among these diseases is an absence of clinical possibilities to stop (to modify) or to reverse (to replace the lost functions of) the disease. The lack of effective treatment creates a tremendous burden on society – the estimated annual economic cost of neurological diseases in Europe is €139 billion, and $180 billion in the USA (WHO, 2006).

To get insights into the mechanisms of neurodegenerative diseases, we need to understand why and how neurons die. Although each neurodegenerative disease has its characteristic affected brain region(s) and exhibits unique symptoms and neuronal pathology, each such disease might share some common mechanisms of neuronal loss on a sub-cellular level. These similarities, once well understood, might represent powerful therapeutic targets. Thus, despite different aetiologies and the as-yet largely unknown nature of neuronal cell death, MND and PD share protein aggregation as common pathology. Protein aggregation in the neuronal cytoplasm and in the axon creates obstacles for a proper connection within the neuronal cell. Neurons especially rely on the transportation of different molecules between the cell body and the extension terminals – and the abrogation of this tightly regulated process would lead to the dying of neuronal extensions (neurodegeneration). As described below in 'Pathomechanisms of neurodegeneration discovered using iPSC', axonal transport and its alterations in the presence of protein aggregates can be nicely followed in human disease-specific neurons using stem cell technology.

An important aspect of the investigation into neurodegeneration mechanisms in the search for new treatment is the experimental model used. Since human neurons are not accessible from a living organism, most of the knowledge about the mechanisms of neuronal pathogenesis discovered over the last few decades has been derived from cell cultures and animal models. However, the relation of those mechanisms to the pathological processes in the human brain is still in debate. On the other side, post-mortem material from patients does allow for the investigation of pathology in human tissue, although it represents the end-stage of the disease. The possibility of analysing neurodegenerative pathways directly in human neurons would provide a more suitable model for studying human neurodegenerative diseases and would give the opportunity to test different approaches to stop or reverse pathology. The recently developed methodology of obtaining iPSC from human somatic cells has revolutionized neurobiological research. The beauty of this approach lies in using easily accessible human cells, such as skin cells, to obtain difficult-to-get cells, such as brain cells. The iPSC technology, thus, provides access to patient-specific brain cells, which can be cultured in a dish and used for the investigation of pathological mechanisms of the disease (Park et al., 2008). The usage of iPSC technology in discovering mechanisms of neuronal damage in MND and PD will be discussed in more detail in 'Pathomechanisms of neurodegeneration discovered using iPSC'. In the long term, the aim is not only to discover pathological mechanisms underlying human neurodegeneration, but also to be able to use this knowledge for developing new, effective individualized treatments.

iPSC as a modelling tool for human neurodegeneration

Taking into account the existing difficulties in the development of new therapeutic approaches for neurodegenerative disorders, and in the discovery of disease-relevant pathological mechanisms in human neurons, it is not surprising that stem cell technology is gaining special attention in the investigation and treatment of neurodegenerative diseases. Human iPSC, since their discovery by the revolutionary work of Shinya Yamanaka and colleagues (Takahashi and Yamanaka, 2006; Takahashi et al., 2007), have been widely used by researchers of human diseases including neurodegeneration. As mentioned above, due to their pluripotent nature, iPSC, derived in an artificial way (called reprogramming) from the adult somatic cells of a patient, can become any somatic cell of the human body – this allows access to patient-specific and disease-relevant cells.

Patient-specific iPSC reprogramming

Technically, iPSC are usually obtained from human skin cells (fibroblasts). Fibroblasts are gained from skin biopsies – a small piece of skin 4 mm in diameter taken from the upper arm. Before taking a biopsy, the patient suffering from a neurodegenerative disease is seen by a neurologist who performs relevant careful clinical examinations. In the outpatient clinic for movement disorders in the Department of Molecular Neurology at the University Hospital of Erlangen led by Prof. Jürgen Winkler, 900 patients with 350 different neurodegenerative diseases per year are seen. Patients are seen twice a year and receive follow-up checks years thereafter, thereby providing a full and detailed clinical history and long-term follow-up data which can be usefully correlated with laboratory results. After the biopsy is taken, the sample is processed further in the laboratory. To ensure anonymity, all patients' data are encoded.

In the laboratory, skin biopsy samples are mechanically and enzymatically digested to obtain single cells and to establish a fibroblast cell culture. Each fibroblast culture is very precisely documented and a part of the fibroblast culture is stored for a longer time (cryopreserved) using liquid nitrogen at −196°C. Expanded fibroblasts can be used for reprogramming. 'Classical' reprogramming is achieved through the introduction of four factors – OCT4, KLF4, SOX2, c-MYC (also called 'Yamanaka factors') – into fibroblasts (Yamanaka, 2007) (also described in Chapter 5, in this volume). These reprogramming factors switch on the pluripotency machinery in fibroblasts that is self-regulating and maintains the stem-cell fate further. Two weeks later, iPSC clones start to appear and need to be manually separated from non-reprogrammed fibroblasts. All iPSC clones will be routinely controlled for their quality[1] and they can also be cryopreserved. Only iPSC clones passing all quality control tests are considered for further investigation.

Obtaining disease-relevant patient-specific neurons from iPSC

In the next step, iPSC can be differentiated into organ-specific cells such as neural cells. The protocol of obtaining disease-specific neurons from iPSC mimics the embryonic development stages of the nervous system (Koch et al., 2009) via the stage of neural precursor cells (NPC).[2] NPC can be expanded, cryopreserved, or further differentiated either into neurons or into other brain cells (glial cells) using an addition of factors important for their establishment during embryonic development. Taking into account the large variety of neurodegenerative diseases and types of affected neurons, it is especially important to obtain disease-relevant types of neurons for experiments or for potential clinical use. For example, dopaminergic neurons are of great interest to PD research, while corticospinal or alpha motor neurons need to be investigated from MND patients. Optimization for obtaining specific neuronal types is performed in the laboratory dish by using factors known to be important for the development of a brain region, where the desired neuronal type is located (Karumbayaram et al., 2009; Reinhardt et al., 2013). Final neurons will be tested for their type by the presence of specific markers (molecules that can be found only in this neuronal type) to characterize the efficiency of a given procedure to obtain neurons from iPSC of a specific type. Whereas it is possible to enhance a yield of some neuronal subtypes, there is still a lack of well-defined laboratory methods for obtaining a lot of specific neuronal subtypes and for obtaining pure cultures of a specific subtype of interest. This part requires further investigation and experimental optimization trials.

Routinely, from each individual, around ten iPSC clones are generated and two to three qualitative iPSC clones will be further used to generate NPC. Normally, several NPC lines are generated out of one iPSC clone; and several parallel rounds of obtaining neuronal cells need to be performed from each NPC line. This complicated scheme is necessary to control for technical variability and for obtaining reliable and reproducible experimental results. Precise nomenclature is needed in order to keep track of the original iPSC clones, NPC lines, and their cellular derivate. It takes roughly three months to start with human fibroblasts and to end up with neuronal culture. It thus requires a careful and precise way of working in order not to lose a long-term culture and to obtain reliable results. Once the neurons are produced from patient-specific iPSC, they can be used for disease modelling. As stated previously, well-established models of diseases are essential to decipher the pathological mechanisms initiating the disease or contributing to the disease progression.

Pathomechanisms of neurodegeneration discovered using iPSC

Having described what neurodegenerative diseases are, and how their treatment may be enhanced by emerging iPSC technology, it will now be shown how such technology has already been used to discover some of the pathomechanisms of MND and PD – a discovery that presents potential therapeutic targets.

Because obtaining disease-specific cells, such as neurons and glial cells, from a living patient is difficult, cellular and animal models are widely used to investigate neurodegeneration. However, animal and cellular models do not fully recapitulate all aspects of human disease and it is obvious that large differences exist between humans and the other animals. Therefore, the mechanisms discovered in animal models, although providing valuable insights into disease pathology, need to be proved in the human system. The discovery of iPSC and the possibility of differentiating them into neurons opened a new era for neurodegenerative research, allowing for the investigation of pathomechanisms specifically in human disease-related cells (Park et al., 2008). Of course, work with patient-specific material and data requires tight ethical regulation; therefore it is essential to apply for ethical approval for each experimental project involving the investigation of human iPSC and derivatives. However, using iPSC for research projects is advantageous compared to embryonic stem cells (ESC) in ethical terms, since no human embryos are needed.

iPSC-based discoveries in PD neurons

Several important pathomechanisms in HSP- and PD-related neurons have been discovered using iPSC-based modelling. Concerning PD pathology, animal models suggest a pathological protein clumping followed by the degeneration of neuronal extensions in the substantia nigra, which is trigged by small toxic aggregates of the protein alpha-synuclein (Winner et al., 2011). To find out whether this phenomenon is also occurring in PD pathology and what the underlying mechanism is, neurons obtained from human iPSC have been studied. Two different approaches have been applied: neurons derived from iPSC of PD patients; and neurons differentiated from the iPSC of healthy individuals, genetically manipulated to contain small pathological alpha-synuclein aggregates to mimic PD pathology. Next, those neurons were cultured in a specific way, allowing the parallel growth of axons and the microscopic observation of the transportation of specialized power-producing cellular organelles – mitochondria – within axons. The transportation of mitochondria between the neuronal body and the axonal end is crucial for neuronal function and survival. In human iPSC-derived neurons, it could be demonstrated for the first time that small alpha-synuclein aggregates disrupt mitochondria transportation within axons and cause axon degeneration and the loss of neuronal connections called synapses. Once synapses are lost, neuronal connectivity is severely affected. Precisely this loss of neuronal connectivity between substantia nigra and the striatum is also responsible for the motor symptoms in PD as described above. Thus, applying iPSC-based PD modelling, it was possible to follow the PD-related pathological phenotype first observed in a PD animal model and, moreover, to identify mechanisms triggering this phenotype.

Disease mechanisms of HSP, determined in iPSC-derived neurons

To investigate the pathomechanisms of HSP, where axonopathy of upper motor neurons occurs, comparisons were made between neurons derived from patients with the most frequent form of HSP caused by mutations in the SPG4, or spastin, and healthy individuals (Havlicek et al., 2014). Morphological analysis revealed the less complex nature of SPG4 neurons with significantly shorter and less branched extensions compared to healthy neurons. Moreover, neuronal extensions of SPG4 neurons showed extensive swellings with a disrupted structure of microtubules, which are an important component of the cellular skeleton and build the roads for the axonal transport of molecules and organelles. The neuronal phenotype observed in the laboratory bears a good resemblance to the disease situation, where patients with mutated SPG4 suffer from progressive spasticity and paresis of the lower limbs due to the degeneration of motor neurons. Importantly, the pathological phenotypes observed in SPG4 neurons were fully reverted by a genetically introduced functional SPG4 protein into these neurons, confirming a gene-dosage dependent neurodegenerative effect of spastin in human SPG4 neurons. Furthermore, these results not only reveal important phenotypes and mechanisms of SPG4-related HSP but they also suggest that expressing functional spastin protein to a physiological level might be an opportunity to halt neuronal degeneration in patients. This, in turn, opens a platform for testing pharmacological compounds for the ability to increase spastin expression, which might help to develop new therapeutics for HSP.

Does iPSC-based research give us a hope for curing neurodegeneration?

As mentioned in the previous paragraphs, deep knowledge of the mechanisms driving human neurodegeneration could create the proper background for the development of new reliable diagnostic tests or for more effective treatments of neurodegenerative diseases. Disease mechanisms can be more easily discovered if one has a good disease model that provides knowledge about early markers and/or early disease pathways. The best model for investigating human disease mechanisms would be a human model. However, human material (except blood) affected by disease is difficult or impossible to obtain from a living human, especially with regard to disorders of the nervous system. With the establishment of iPSC technology, neurological research has received a significant boost as it opens new avenues to obtain patient-specific, disease-relevant human cells and to utilize those for modelling disease processes (Park et al., 2008). It also gives researchers the possibility of following the disease in a relevant patient-specific human system from the very early stage over its progression up to the late stage. In this section, we try to think about the iPSC potency for the treatment of neurodegenerative diseases.

Clinical advantages and disadvantages of iPSC

At present, iPSC technology is intensively used for investigating disease mechanisms. Over the past years, numerous groups have revealed disease-specific phenotypes in models based on human iPSC technology for a number of neurodegenerative diseases as well as for neuropsychological conditions, peripheral neuropathies, cardiac syndromes, and premature ageing (Brennand et al., 2011; Lee et al., 2009; Liu et al., 2011; Marchetto et al., 2010; Moretti et al., 2010). These studies proved the concept of successful disease modelling using patient-specific human iPSC. Moreover, so-called 'personalized medicine' may become more realistic as access to individual-specific cells becomes viable, with the possibility of testing individual disease progress, cell susceptibility to stressors, and response to treatments. Furthermore, iPSC-derived patient-specific cells might in the future provide novel sources for autologous cell replacement therapies (Jung et al., 2012; Yamanaka, 2007). On one side, the autologous nature of iPSC cellular derivates would reduce the risk of transplant rejections by the host tissue. On the other side, however, genetic manipulations and the laboratory culture required to generate and expand iPSC potentially increase the possibility of tumour formation after transplantation into patients, which is a great concern (Miura et al., 2009). Therefore, the careful and extensive quality examination of patient iPSC lines and their derivates must be performed in order to enhance the feasibility of iPSC-derived cells for transplantation therapies. It is worth mentioning that fibroblasts and human cord blood stem cells have also been converted into neurons (Ambasudhan et al., 2011; Jung et al., 2012). A direct conversion of somatic adult cells into another type of somatic cell could minimize the embryo-related ethic debate.

New strategies for iPSC application

As stated above, significant discoveries of disease-causing and disease-driving mechanisms have been made in neurons obtained from patient-specific iPSC. It is however important to note that neurons, even being disease-specific, might behave differently in a dish in contrast to a diseased environment, such as the brain tissue of a patient with neurodegenerative disease. This could limit the wide extrapolation of the results obtained with iPSC-derived cells. To overcome these difficulties, scientists have begun to produce iPSC-derived three-dimensional models of specialized tissues, called organoids, which have key features of their counterparts in a living organism. Organoids are useful systems for investigating a cell in a physiological and/or diseased surrounding. For example, cerebral organoids called 'mini brains' have been used to model the complex neurodevelopmental disorder microcephaly that causes abnormal growth of the brain (Clevers, 2016).

With the fast progress in stem cell technologies recently, there is an increasing need of research consortia that comprise experts from different fields, including those of science, the clinic, industry, ethics, and even comprising patients, in order to revolutionize the application and clinical translation of these recent developments. Some such consortia already exist. For example, the research-orientated Bavarian regional consortium 'ForIPS' aims to investigate the pathology of PD using human iPSC-based modelling of the disease.[3] The more clinically-focussed European consortium 'TRANSEURO' is orientated towards developing efficacious and safe foetal cell-based treatments for PD.[4] The joined forces within such consortia profoundly stimulate further research and its application in the clinic.

Concluding remarks

The complex nature and pathology of human neurodegenerative diseases require deeper understanding in order to be effectively treated. Stem cell technology based on iPSC has revolutionized neurodegenerative research due to its potential to provide researchers with patient-specific material. Starting from the somatic cells of a patient (which are usually easy to obtain), they are reprogrammed to pluripotent stem cells that can be, in turn, used to obtain various somatic cells (which are difficult to access from living individuals, for example nerve cells). iPSC-based disease-related neurons or brain cells can be used to study the disease mechanisms in the human system, overcoming the limitations, on the one hand, of using animal models, which do not fully recapitulate human pathology, and, on the other hand, overcoming ethical problems connected to the use of human embryos. Moreover, iPSC-based disease models offer perfect possibilities for translational research: 1) they serve as a valuable platform for pre-clinical compound screens with the goal of discovering new drug candidates for disease treatment, (2) they promote attempts to reverse the disease by the transplantation of lost brain cells, and (3) they inspire the development of individualized medicine. Thus, iPSC technology provides new hope for many patients with hitherto incurable diseases, as well as for their physicians in their ability to efficiently treat or even cure these diseases one day.

Notes

1 Quality control includes morphological, pluripotency, and genetic testing. Morphologically, iPSC have a high nucleus-to-cytoplasm ratio, and grow in very dense, flat colonies with rather smooth, round borders. The pluripotency criteria of iPSC are the expression of endogenous pluripotency-associated markers such as Nanog, OCT4, Tra1-60, Tra1-81, Lin28, and their ability to develop cells of the three embryonic germ layers by either teratoma assays in immuno-deficient mice or differentiation assays in cell culture. Genetically, karyotype analysis is performed to identify iPSC clones without chromosomal abnormalities (which can be acquired during culture).
2 The NPC quality needs to be well controlled by the expression of the NPC-specific transcription factors such as Nestin and Sox2. Only NPC cultures consisting of more than 90% of Nestin/Sox2–double positive cells can be used as a source for reasonable neuronal differentiation.

3 www.bayfor.org/en/portfolio/research-cooperations/world-of-living/forips. html.
4 www.transeuro.org.uk.

References

Ambasudhan, R., Talantova, M., Coleman, R., Yuan, X., Zhu, S., Lipton, S. A. and Ding, S. (2011). Direct reprogramming of adult human fibroblasts to functional neurons under defined conditions. *Cell Stem Cell*, 9(2), pp. 113–118.

Blackstone, C. (2012). Cellular pathways of hereditary spastic paraplegia. *Annual Review of Neuroscience*, 35, pp. 25–47.

Blackstone, C., O'Kane, C. J. and Reid, E. (2011). Hereditary spastic paraplegias: Membrane traffic and the motor pathway. *Nature Reviews Neuroscience*, 12(1), pp. 31–42.

Brennand, K. J., Simone, A., Jou, J., Gelboin-Burkhart, C., Tran, N., Sangar, S., Li, Y., Mu, Y., Chen, G., Yu, D., McCarthy, S., Sebat, J. and Gage, F. H. (2011). Modelling schizophrenia using human induced pluripotent stem cells. *Nature*, 473(7346), pp. 221–225.

Clevers, H. (2016). Modeling development and disease with organoids. *Cell*, 165(7), pp. 1586–1597.

Connolly, B. S. and Lang, A. E. (2014). Pharmacological treatment of Parkinson disease: A review. *JAMA*, 311(16), pp. 1670–1683.

Fink, J. K. (2013). Hereditary spastic paraplegia: Clinico-pathologic features and emerging molecular mechanisms. *Acta Neuropathologica*, 126(3), pp. 307–328.

Havlicek, S., Kohl, Z., Mishra, H. K., Prots, I., Eberhardt, E., Denguir, N., Wend, H., Plotz, S., Boyer, L., Marchetto, M. C., Aigner, S., Sticht, H., Groemer, T. W., Hehr, U., Lampert, A., Schlotzer-Schrehardt, U., Winkler, J., Gage, F. H. and Winner, B. (2014). Gene dosage-dependent rescue of HSP neurite defects in SPG4 patients' neurons. *Human Molecular Genetics*, 23(10), pp. 2527–2541.

Hickey, P. and Stacy, M. (2016). Deep brain stimulation: A paradigm shifting approach to treat Parkinson's disease. *Frontiers in Neuroscience*, 10, p. 173.

Huse, D. M., Schulman, K., Orsini, L., Castelli-Haley, J., Kennedy, S. and Lenhart, G. (2005). Burden of illness in Parkinson's disease. *Journal of Movement Disorders*, 20(11), pp. 1449–1454.

Jung, Y. W., Hysolli, E., Kim, K. Y., Tanaka, Y. and Park, I. H. (2012). Human induced pluripotent stem cells and neurodegenerative disease: Prospects for novel therapies. *Current Opinion in Neurology*, 25(2), pp. 125–130.

Kalia, L. V. and Lang, A. E. (2015). Parkinson's disease. *Lancet*, 386(9996), pp. 896–912.

Kandel, E. R. and Squire, L. R. (2000). Neuroscience: Breaking down scientific barriers to the study of brain and mind. *Science*, 290(5494), pp. 1113–1120.

Karumbayaram, S., Novitch, B. G., Patterson, M., Umbach, J. A., Richter, L., Lindgren, A., Conway, A., Clark, A., Goldman, S. A., Plath, K., Wiedau-Pazos, M., Kornblum, H. I. and Lowry, W. E. (2009). Directed differentiation of human induced pluripotent stem cells generates active motor neurons. *Stem Cells*, 27(4), pp. 806–811.

Klein, C. and Westenberger, A. (2012). Genetics of Parkinson's disease. *Cold Spring Harbor Perspectives in Medicine*, 2(1), p. a008888.

Koch, P., Opitz, T., Steinbeck, J. A., Ladewig, J. and Brustle, O. (2009). A rosette-type, self-renewing human ES cell-derived neural stem cell with potential for in vitro instruction and synaptic integration. *Proceedings of the National Academy of Sciences USA*, 106(9), pp. 3225–3230.

Lee, G., Papapetrou, E. P., Kim, H., Chambers, S. M., Tomishima, M. J., Fasano, C. A., Ganat, Y. M., Menon, J., Shimizu, F., Viale, A., Tabar, V., Sadelain, M. and Studer, L. (2009). Modelling pathogenesis and treatment of familial dysautonomia using patient-specific iPSCs. *Nature*, 461(7262), pp. 402–406.

Liu, G. H., Barkho, B. Z., Ruiz, S., Diep, D., Qu, J., Yang, S. L., Panopoulos, A. D., Suzuki, K., Kurian, L., Walsh, C., Thompson, J., Boue, S., Fung, H. L., Sancho-Martinez, I., Zhang, K., Yates, J., 3rd and Izpisua Belmonte, J. C. (2011). Recapitulation of premature ageing with iPSCs from Hutchinson-Gilford progeria syndrome. *Nature*, 472(7342), pp. 221–225.

Marchetto, M. C., Carromeu, C., Acab, A., Yu, D., Yeo, G. W., Mu, Y., Chen, G., Gage, F. H. and Muotri, A. R. (2010). A model for neural development and treatment of Rett syndrome using human induced pluripotent stem cells. *Cell*, 143(4), pp. 527–539.

Miura, K., Okada, Y., Aoi, T., Okada, A., Takahashi, K., Okita, K., Nakagawa, M., Koyanagi, M., Tanabe, K., Ohnuki, M., Ogawa, D., Ikeda, E., Okano, H. and Yamanaka, S. (2009). Variation in the safety of induced pluripotent stem cell lines. *Nature Biotechnology*, 27(8), pp. 743–745.

Moretti, A., Bellin, M., Welling, A., Jung, C. B., Lam, J. T., Bott-Flugel, L., Dorn, T., Goedel, A., Hohnke, C., Hofmann, F., Seyfarth, M., Sinnecker, D., Schomig, A. and Laugwitz, K. L. (2010). Patient-specific induced pluripotent stem-cell models for long-QT syndrome. *The New England Journal of Medicine*, 363(15), pp. 1397–409.

Novarino, G., Fenstermaker, A. G., Zaki, M. S., Hofree, M., Silhavy, J. L., Heiberg, A. D., Abdellateef, M., Rosti, B., Scott, E., Mansour, L., Masri, A., Kayserili, H., Al-Aama, J. Y., Abdel-Salam, G. M., Karminejad, A., Kara, M., Kara, B., Bozorgmehri, B., Ben-Omran, T., Mojahedi, F., Mahmoud, I. G., Bouslam, N., Bouhouche, A., Benomar, A., Hanein, S., Raymond, L., Forlani, S., Mascaro, M., Selim, L., Shehata, N., Al-Allawi, N., Bindu, P. S., Azam, M., Gunel, M., Caglayan, A., Bilguvar, K., Tolun, A., Issa, M. Y., Schroth, J., Spencer, E. G., Rosti, R. O., Akizu, N., Vaux, K. K., Johansen, A., Koh, A. A., Megahed, H., Durr, A., Brice, A., Stevanin, G., Gabriel, S. B., Ideker, T. and Gleeson, J. G. (2014). Exome sequencing links corticospinal motor neuron disease to common neurodegenerative disorders. *Science*, 343(6170), pp. 506–511.

Park, I. H., Arora, N., Huo, H., Maherali, N., Ahfeldt, T., Shimamura, A., Lensch, M. W., Cowan, C., Hochedlinger, K. and Daley, G. Q. (2008). Disease-specific induced pluripotent stem cells. *Cell*, 134(5), pp. 877–886.

Reinhardt, P., Glatza, M., Hemmer, K., Tsytsyura, Y., Thiel, C. S., Hoing, S., Moritz, S., Parga, J. A., Wagner, L., Bruder, J. M., Wu, G., Schmid, B., Ropke, A., Klingauf, J., Schwamborn, J. C., Gasser, T., Scholer, H. R. and Sterneckert, J. (2013). Derivation and expansion using only small molecules of human neural progenitors for neurodegenerative disease modeling. *PLoS One*, 8(3), p. e59252.

Sheikh, S., Safia, Haque, E. and Mir, S. S. (2013). Neurodegenerative diseases: Multifactorial conformational diseases and their therapeutic interventions. *Journal of Neurodegenerative Diseases*, Article ID 563481, p. 8.

Soderblom, C. and Blackstone, C. (2006). Traffic accidents: Molecular genetic insights into the pathogenesis of the hereditary spastic paraplegias. *Pharmacol Ther*, 109(1–2), pp. 42–56.

Spillantini, M. G., Schmidt, M. L., Lee, V. M., Trojanowski, J. Q., Jakes, R. and Goedert, M. (1997). Alpha-synuclein in Lewy bodies. *Nature*, 388(6645), pp. 839–840.

Takahashi, K. and Yamanaka, S. (2006). Induction of pluripotent stem cells from mouse embryonic and adult fibroblast cultures by defined factors. *Cell*, 126(4), pp. 663–676.

Takahashi, K., Tanabe, K., Ohnuki, M., Narita, M., Ichisaka, T., Tomoda, K. and Yamanaka, S. (2007). Induction of pluripotent stem cells from adult human fibroblasts by defined factors. *Cell*, 131(5), pp. 861–872.

WHO (2006). *Neurological Disorders: Public Health Challenges*. Geneva: WHO Press.

Winner, B., Jappelli, R., Maji, S. K., Desplats, P. A., Boyer, L., Aigner, S., Hetzer, C., Loher, T., Vilar, M., Campioni, S., Tzitzilonis, C., Soragni, A., Jessberger, S., Mira, H., Consiglio, A., Pham, E., Masliah, E., Gage, F. H. and Riek, R. (2011). In vivo demonstration that alpha-synuclein oligomers are toxic. *Proceedings of the National Academy of Sciences USA*, 108(10), pp. 4194–9.

Worms, P. M. (2001). The epidemiology of motor neuron diseases: A review of recent studies. *Journal of the Neurological Sciences*, 191(1–2), pp. 3–9.

Yamanaka, S. (2007). Strategies and new developments in the generation of patient-specific pluripotent stem cells. *Cell Stem Cell*, 1(1), pp. 39–49.

8 Stem cell biology

A conceptual overview

Melinda Bonnie Fagan

Abstract

This chapter provides an overview of stem cell biology, through the lens of its central concept ('the stem cell') and experimental practices. I begin by presenting a minimal stem cell model, identifying its key elements and showing how these relate to other core biological concepts. Next, I show how this minimal model is specified in different ways, corresponding to the main varieties of stem cell and major strands of stem cell research. This provides an accessible overview of stem cell research in its present-day configuration. I then discuss several implications of this 'top-down' abstract model, as it relates to experimental practices in stem cell research. One is 'experimental relativity' – stem cells, as experimentally observable biological entities, are relative to particular experimental methods and contexts. Furthermore, stem cells can only be experimentally individuated – i.e. picked out as distinct biological entities – conditional on hypotheses stating that tested stem cell populations are homogeneous. Stem cell concepts also involve substantive assumptions about biological development at organismal, cellular, and molecular levels. At the cellular level, these assumptions can be precisely expressed and analysed as lineage tree models. Multi-level models of development, relating molecular, cellular, and organismal levels, offer a look ahead at emerging explanations of stem cell phenomena. I conclude with a brief discussion of how these incipient explanatory models, and other results of stem cell research, bear on certain ethical debates.

Introduction

Much ethical debate over stem cells concerns the permissibility of using human embryos for research. In the background of these debates are views about biological development: what a developing entity is, the meaning of its 'potential', and what controls the transformation from fertilized egg to whole organism. Stem cell biology aims to clarify all these ideas. So understanding stem cell research is crucial for making sense of the ethical debates. The field's central concept, that of a stem cell, is peculiar in that it unites two very different ideas. A *cell* is a well-characterized biological entity, observable via relatively simple technology and clearly distinguished from its environment and other cells by a bounding membrane. A *stem* is the beginning of a process, the point of origin for something that is to be. A stem cell, then, is both entity and process; a cell defined by what it gives rise to rather than its observable traits. More precisely, a stem cell is generally defined today as a cell that has 'the capacity to both self-renew and give rise to differentiated cells' (Ramelho-Santos and Willenbring, 2007, 35).[1] Cell differentiation is one key aspect of organismal development. So the very idea of a stem cell involves assumptions about the process of development, which often remain tacit and unquestioned. One goal of this chapter is to make these assumptions explicit. To do so I present a minimal definition of 'stem cell', which further explicates the prevailing scientific definition citing self-renewal and differentiation.

This minimal definition, or *model*, accomplishes three things.[2] Firstly, as just noted, it makes explicit certain background assumptions about the nature of development as a process. Secondly, it offers a point of entry into the complex field of stem cell research, which is characterized by daunting technical terminology, rapid change, and multiple meanings of 'stem cell'. The main contours and strands of stem cell research can be understood as different specifications of the same minimal model. So this framing makes possible a kind of overview, which is not easily obtained from reading the scientific literature on stem cells. Thirdly, the minimal model sheds new light on a number of philosophically and scientifically important questions. How are stem cells identified in practice? What assumptions are needed to infer stem cells from experimental data? What consequences do these assumptions and practices have for the nature of stem cells and our understanding of biological development? Although these questions are primarily methodological and conceptual, they bear on the social and practical questions central to this volume.

This chapter provides an overview of stem cell biology through the lens of its central concept: 'the stem cell'. I begin by presenting the minimal stem cell model, identifying its key elements and showing how these relate to other core biological concepts. Next, I show how the minimal model is specified in different ways, corresponding to the main varieties of stem cell and major strands of stem cell research. This provides an accessible overview of stem cell research in its present-day configuration. I then discuss several implications of this 'top-down' abstract model, as it relates to experimental practices in stem cell research. One is 'experimental relativity' – stem cells, as experimentally observable biological entities, are *relative* to particular experimental methods and contexts. Furthermore, stem cells can only be experimentally individuated – i.e. picked out as distinct biological entities – conditional on hypotheses stating that tested stem cell populations are homogeneous. Stem cell concepts also involve substantive assumptions about biological development at organismal, cellular, and molecular levels. At the cellular level, these assumptions can be precisely expressed and analysed as lineage tree models. Multi-level models of development – relating molecular, cellular, and organismal levels – offer a look ahead at emerging explanations of stem cell phenomena. I conclude with a brief discussion of how these incipient explanatory models, and other results of stem cell research, bear on the ethical debates over embryonic stem cells.

Minimal stem cell model

I begin by presenting the minimal stem cell model which explicates the general functional definition of 'stem cell'. In the scientific community today there is clear consensus on this general definition. Douglas Melton's statement in the most recent edition of *Essentials of Stem Cell Biology* is a representative example:

> Stem cells are functionally defined as having the capacity to self-renew and the ability to generate differentiated cells.
>
> (Melton, 2013, 7)[3]

Similar definitions are provided by the International Society for Stem Cell Research and the European Stem Cell Network, two of the main organizations for stem cell researchers worldwide:[4]

> Stem cells: Cells that have both the capacity to self-renew (make more stem cells by cell division) as well as to differentiate into mature, specialized cells.
>
> (ISSCR, 2016)

> Stem cell: a cell that can continuously produce unaltered daughters and also has the ability to produce daughter cells that have different, more restricted properties.
>
> (European Stem Cell Network, 2016)

All these definitions highlight two abilities of stem cells: self-renewal, production of more stem cells by cell division, and differentiation, production of more specialized cells representing later stages of development. To further clarify the stem cell concept we need to further analyse these two reproductive processes.

The ability to self-renew is realized by cell division. A cell that has the ability to self-renew has the ability to divide to produce an offspring cell that resembles itself. A brief review of cell theory is useful here. Its basic tenets, established in the mid-nineteenth century, still hold:[5]

i   Cells reproduce by binary division; a parent cell divides to produce two offspring cells.
ii An individual cell's existence begins with a cell division event and ends with either a second division event (producing two offspring) or cell death (and no offspring).[6]
iii Generations of cells linked by reproductive division form a lineage.

A cell lineage is a biological entity composed of successive cell generations, organized by reproductive relations. So a stem cell is not simply one individual cell, existing between division events. The stem cell concept

also implies the notion of a cell lineage: individual cells related by reproduction (Figure 8.1). This is implicit in the very term, 'self-renewal', which
suggests that something persists in cell reproduction. But this something
is not a cell; a parent cell ceases to exist upon dividing. What persists is
the *cell lineage*: a biological entity extending over multiple generations.
Often, when people speak of 'stem cells', what they mean is the stem cell
lineage, or *line*. A self-renewing stem cell gives rise to a continuous lineage of stem cells.

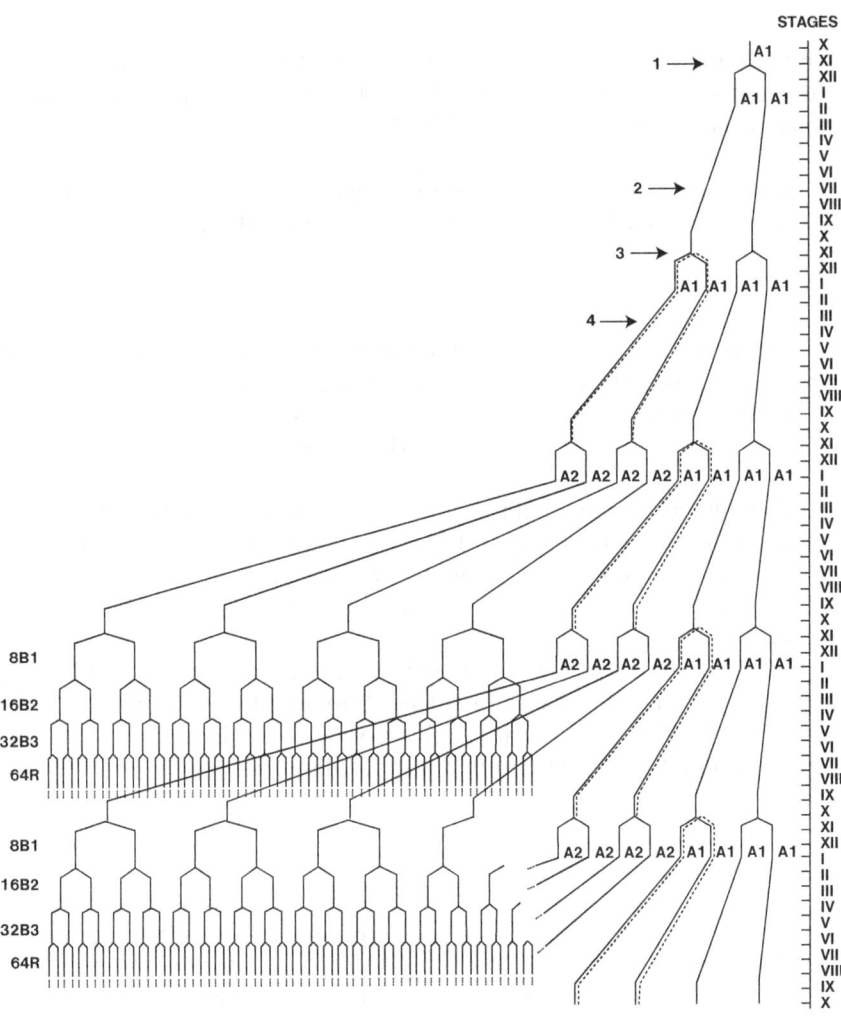

Figure 8.1 Cell lineage diagram.

Source: Clermont and Leblond 1959, 267; Copyright © 1959 Wiley-Liss, Inc. Reprinted by
permission from Wiley.

The concept of self-renewal implies not only a cell lineage but also a comparison across its generations. A cell possessing the ability to self-renew has the ability to divide to produce one or two cells that resemble it.[7] Because no two numerically distinct cells are the same in absolutely *every* respect, a scientifically useful notion of comparison across cell generations must presume a set of variable characters relative to which such comparisons are made such as cell size, shape, or expression level of a particular gene. For a stem cell, of course, self-renewal means production of another stem cell like the parent. So the characters to be compared across generations are just those taken to define the stem cell of interest.

The concept of differentiation brings further complexities to the notion of a stem cell. The ability to differentiate is not necessarily realized by cell division: an individual cell may differentiate by transforming directly from one 'cell state' to another. A cell state, in current parlance, is a pattern of gene expression and molecular interactions that determines a cell's structural and functional characteristics. So the idea of a cell state is a multi-level concept, linking cell and molecular levels of biological organization (see below). Self-renewal is cell division without a change in cell state. Differentiation is change in cell state, which may or may not be coordinated with cell division. In the former case, one or more offspring cells exhibit a cell state different from that of the parent. Accordingly, scientists often propose the following as a minimal or basic definition of a stem cell:

> A stem cell is a cell that can divide asymmetrically, so one offspring is a stem cell and the other is a more specialized cell (Figure 8.2).

This 'basic definition' has the virtue of simplicity – a rare commodity in stem cell research! The simple model illustrates both self-renewal and differentiation in a single cell division event, which produces one offspring cell resembling the parent and another resembling a more specialized cell type. In practice, however, stem cell researchers are rarely interested in cell lineages of only two generations. The basic definition can be generalized by introducing a variable n, representing the number of cell divisions. Assuming regular rates of cell division, n also serves as a time parameter, ranging from hours to decades of conventional calendar time. So the generalized basic definition is as follows:

> A stem cell is a cell that initiates a cell lineage L, which after n divisions produces both stem cells and more specialized cells.

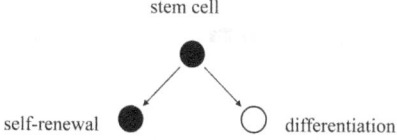

Figure 8.2 Asymmetric cell division illustrating two stem cell capacities.

As we have seen, the cross-generation comparison presupposed in the concept of self-renewal requires a finite set of characters C, relative to which parent and offspring cells are compared.

The idea of cells being more or less specialized requires further explication as well. Differentiation is not simply the converse of self-renewal; a differentiated offspring cell is not merely dissimilar to its parent (with respect to some set of characters C). Rather, the difference is in a particular 'direction'. Here ideas about the nature of development come into play. At minimum, the concept of differentiation presupposes some idea of a developmental process, which provides a basis for ordering cell states as more or less specialized. Importantly, this developmental ordering is conceptually distinct from the process of cell division. An individual cell can traverse multiple cell states, while cells occupying a single state can undergo multiple division events (i.e. self-renewal). The two processes combine in models that represent the sizes and rates of exchange between developmentally ordered 'cell compartments'. Significantly, stem cells were conceived primarily in terms of such compartments during the 1970s–80s, before the field was reoriented by the innovation of cultured embryonic stem cell lines (Potten and Lajtha, 1982; Thomson et al., 1998). The key point is that the stem cell concept presupposes some notion of developmental order for cells.

Putting the above ideas together, the general definition of a stem cell entails the following elements:

- a cell lineage L, consisting of individual cells related by reproduction
- a set of variable characters C, with respect to which cells are compared across generations
- a number of cell divisions n (alternatively, a time interval t)
- a developmental process D, which orders cell states $s_1,...,s_n$ from less to more specialized[8]

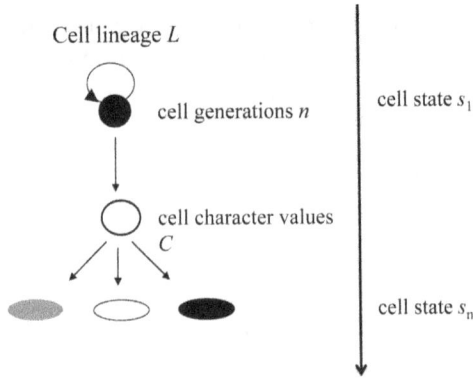

developmental process D

Figure 8.3 Minimal stem cell model, explicating the general definition in terms of self-renewal and differentiation.

These diverse elements can be integrated into a single abstract model, which explicates the general definition of 'stem cell' (Figure 8.3).[9] According to this model, a stem cell is a cell that is the origin of a cell lineage L, generated by n cell divisions and organized by comparison of characters C, which can be mapped onto a developmental process D of ordered cell states $s_1,...,s_n$. Any cell to which values of these variables can be assigned such that this complex condition is satisfied is a cell that is capable of both self-renewal and differentiation, and thus a cell that conforms to the general definition of 'stem cell'.

Overview of stem cell research

The above model does not directly represent biological cells or cell lineages. Relatedly, the general definition of stem cells as capable of self-renewal and differentiation is not used as a set of necessary and sufficient conditions for a cell to be a stem cell. So the abstract model explicating this general definition does not correspond to all and only stem cells that are known to science or yet to be discovered. At this time, there is no such unified account of stem cells. Rather, the general definition is a starting point for more detailed characterizations of particular varieties of stem cell: adult, embryonic, pluripotent, induced, neural, epiblast, hematopoietic (blood-forming), and more. We can use the abstract model to clarify these various types of stem cell, and gain an overview of the field in terms of the main ones under investigation. Briefly, the different varieties of stem cell correspond to different combinations of values of variables L, n, C, D, and $s_1,...,s_n$ (Table 8.1). Of particular interest here are variables L and D, referring to cell lineage and developmental process, respectively.

As presented above, stem cells are defined in the first instance in terms of cellular entities, characters, and processes. But, on closer inspection, whole multicellular organisms are implicated in the definition as well.

Table 8.1 Varieties of Stem Cell Organized in Terms of the Abstract Model

Type	L (source)	C	n	D
ESC	5-d embryo ICM	Cell size, cell shape, gene expression, karyotype, telomerase activity, alk-phos, cell surface molecules	≥50 divs	Traits of cells from three germ layers
HSC	BM, cord, peripheral blood	Cell size, density, light scatter, surface molecules, cell cycle status	>6 months	Traits of main blood and immune cell lineages
NSC	Basal lamina of ventricular zone	Cell morphology, surface markers, gene expression, cytokine response	Months to years	Traits of neurons, astrocytes, and oligodendrocytes

(Continued)

Type	L (source)	C	n	D
iPSC	Various (relatively mature cells)	Colony shape, cell size, cell shape, nucleus/cytoplasm ratio, cell surface molecules, activity and expression of specific proteins, gene expression (specific and global), histone modifications at key locations	≥50 divs	Traits of cells from three germ layers
GSC	5–9-wk gonadal ridge	Colony shape, alk-phos, surface expression (SSEA-1, SSEA-3, SSEA-4, TRA-1–60, TRA-1–81)	20–25 wks	Traits of cells from three germ layers
EC	Teratocarcinoma (129)	Cell shape, morphology, production of embryoid bodies, surface molecules, enzymes	Unlimited	Traits of cells from three germ layers, teratocarcinoma

Most obviously, stem cells are derived from an organismal source. Every stem cell is a part of, or is descended via cell division from a part of, exactly one multicellular organism. The scientific names of major stem cell varieties often refer to features of the source organism – notably, species and developmental stage. For example, 'hESC' is an abbreviation of 'human embryonic stem cell', 'm-epiSC' of 'mouse epiblast stem cell'. Stem cells found within a mature organism ('adult stem cells') are named for the part of that organism they build and/or replenish: neural stem cells, hematopoietic stem cells, epithelial stem cells, muscle stem cells, etc. More broadly, stem cell research has long been divided into two branches, which focus on 'adult' or 'embryonic' stem cells, respectively. The adult/embryonic distinction refers to the developmental stage of stem cells' organismal source.

Accordingly, variable *L* can be further broken down into three subvariables characterizing a stem cell's lineage in terms of its organismal source:

- species
- developmental stage
- location within the organism

The third sub-variable does not denote a feature of an organism, but specifies a *relation* between cell and organism – namely, that the former occupies a particular position within the latter. In principle, stem cell biologists could investigate the full range of values for these three sub-variables, characterizing the properties of stem cells derived from different species, developmental stages, and organismal parts. But in practice the field is less comprehensive and systematic. This is largely because stem cell research is oriented towards medical applications at least as much as towards increased knowledge of cell development. The field's primary goal is to invent new therapies for a wide range of pathological conditions, harnessing stem cells' regenerative capacities. Due to the field's clinical aspirations, most stem cell research is done on just two species: human and mouse – biomedicine's intended beneficiaries and its paradigm model organism. Within these species, the preponderance of research is directed to developmental stages that can give rise to pluripotent stem cells, and to locations within the organism from which stem cells are derived that produce or replenish clinically relevant tissues: blood, neurons, skin, cardiac muscle, etc.

Closely related to the adult/embryonic distinction is the classification of stem cells in terms of 'potency'. Cognates of this term are qualitative descriptions of the range of mature (specialized) cell types that a given stem cell can give rise to by differentiation (accompanied by cell division). Maximum developmental potential is referred to as *totipotency*: the capacity to produce an entire organism (and, in mammals, extra-embryonic tissues). In animals, totipotency is limited to the fertilized egg and products of early cell divisions. In the nineteenth and early twentieth century, the term 'stem cell' often referred to totipotent cells of early development (Dröscher, 2014). But this usage is no longer standard. The maximum developmental potential for stem cells as the term is used today is *pluripotency*: the ability to produce all cell types of an adult organism. More restricted stem cells are *multipotent*: able to produce some, but not all, mature cell types. Stem cells that can give rise to only a few mature cell types are *oligopotent*. Minimum differentiation potential is *unipotency*: the capacity to produce only a single cell type. The rough ordering of 'potencies' offers a generic way of classifying stem cells. Pluripotent stem cells are the focus of much stem cell research, as they can serve as sources of stem cells with more restricted developmental potential. For example, multipotent hematopoietic or neural stem cells can be produced by differentiation from pluripotent stem cells.

The concept of stem cell potency weaves together cell-level and organism-level development in at least two ways. First, a given stem cell's developmental potential is conceptualized in terms of an adult multicellular organism. If stem cell potency were characterized in cellular terms only, the 'pluri-', 'multi-', and 'uni-' monikers would simply refer to 'raw numbers' of cell types. But the designations of potency refer not to cell types considered in isolation, so to speak, but relative to the completed process of organismal development. Pluri-, multi-, and unipotent stem cells are capable of giving rise to all, some, or one of the cell types making up the body of the source organism from which they are derived. So, multicellular organisms are implicated in both the sources of stem cells and their potential products. Furthermore, the ordering of stem cells by potency is thought to be correlated with the developmental stage of the source organism. That is, the earlier the latter's developmental stage, the greater the former's developmental potential (Figure 8.4). As development of a multicellular organism proceeds, the developmental potential of stem cells derived from that organism diminishes. Stem cells derived from an early embryo are pluripotent, while stem cells isolated from parts of a fully developed organism of the same species are (usually) multi-, oligo-, or unipotent. The question of how strict this correspondence is, and whether pluripotent stem cells can be reliably found in adult organisms, is a matter of protracted controversy (see Fagan, 2013a, Chapter 3).

These considerations about stem cell potency suggest further assumptions relevant to developmental process D, which orders cell states on a continuum of differentiation. The simplest way to conceive a developmental process, at the cell level, is as a linear sequence of cell states ordered in time. This is how whole-organism stages of embryonic development were represented, in the early days of the science (Hopwood, 2005). But cell developmental processes are coordinated with cell division events, such that stages are often arranged in a branching hierarchy rather than a single linear sequence. So variable D can, like variable L, be further analysed as a set of sub-variables, which together delimit the range of possible topologies for cell lineage trees:

- d, number of stages (depth of developmental hierarchy)
- p, number of termini (a cell's developmental potential)
- a, arrangement of branch-points (the number of which $= p{-}1$)

In this way, variable D captures not only the characterization of a given stem cell's potency, but also the topological structure of the cell lineage to which it gives rise.[10]

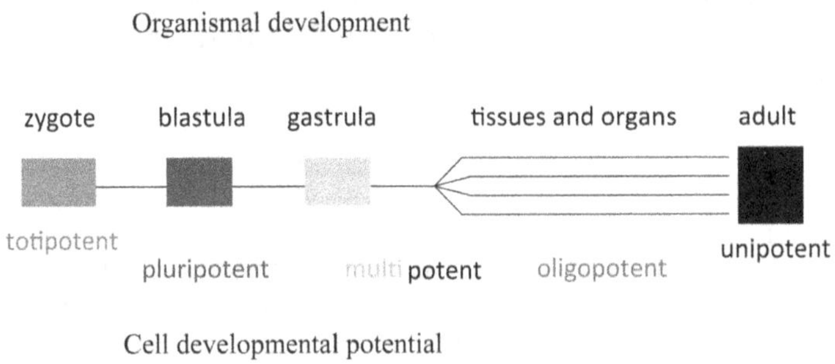

Figure 8.4 Correlated processes of organismal development and restriction of stem cell developmental potential.

Experimental relativity

The remainder of the chapter examines several implications of the model presented above, which bear on current debates in the philosophy of science and the philosophy of biology. The first of these is what I term 'experimental relativity': stem cells can be identified by experiment only relative to particular methods and hypotheses. We currently have no other methods than experiment to identify and characterize stem cells – the field has markedly few predictive theories. So, in practice, the stem cell concept is not one but many, each variant relative to an experimental context. The context-dependence of stem cells as identifiable biological entities follows from the fact that values of variables L, n, and C are specified by experimental materials and methods.[11]

For example, human embryonic stem cell (hESC) lines are the 'stem' of lineages derived from early human embryos (about five-day-old blastocysts) *in vitro*. hESC lines are derived by removing part of the early embryo's inner layer and placing it in a new artificial cell culture. These details of the experimental method specify sub-variables for L: species, developmental stage, and location within the source organism. In their new environment, some of the cultured cells divide to produce colonies with specific morphological and molecular characteristics, including rapid division in culture, lack of specialized traits, flat round shape, large nuclei surrounded by correlatively thin cytoplasm, prominent nucleoli, high telomerase activity, and high expression of particular genes (Thomson et al., 1998). These details of the method specify values for the set of cell characters C. Cells with these features are selected for further *in vitro* culture, dividing to generate new colonies. Continuing cycles of colony formation and selection yield an 'immortal' cell line. The requirement for continued cell division entails that there is no upper limit for variable n. So cells resulting from this procedure are automatically self-renewing. Their differentiation potential is tested by moving them to environments conducive to differentiation, and comparing results with characters of ordinary differentiated cells. These details of the method partly specify the values of variable D – the traits that distinguish more from less differentiated cell states.

Other kinds of stem cell are identified by different experimental procedures, corresponding to different combinations of values of the variables in the abstract model. Although their specific details differ, experiments that aim to identify stem cells have a basic structure of three stages (Figure 8.5). In the first, cells are extracted from an organismal source (specifying L) and placed in a new environment in which candidate stem cell characters C are measured. The duration of this part of the experiment specifies the value of n (time or cycles of self-renewal). In the second stage, measured cells are moved to another environment in which capacities for differentiation can be realized. Finally, characters of differentiated cells (the termini of D) are measured. So details of the experimental procedure fix values of L, n, C,

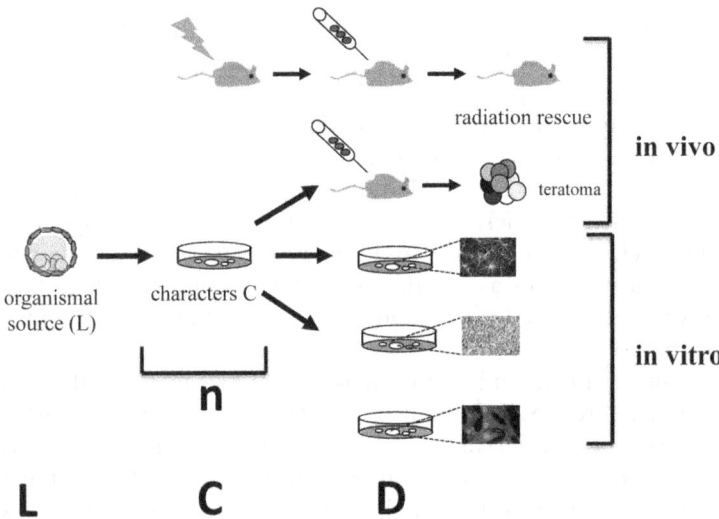

Figure 8.5 Basic design of stem cell experiments, with stages indicating how values of variables in the minimal model are determined.

and (partly) *D*. Experiments with this basic design are the only way stem cells are identified, in practice. In this sense, stem cells are relative to the experimental context in which their characteristic abilities are determined. A consequence of this relativity is that, without knowledge of the experimental context, claims about stem cells are massively ambiguous. If we do not know the experimental details that fix values of the abstract model's variables, we cannot know what the term 'stem cell' refers to in any particular case.

Biological individuals

A related consequence of the minimal model, in conjunction with the design of stem cell experiments, is that stem cells cannot, strictly speaking, be identified at the single-cell level. This result bears on debates about biological individuality, as well as evidential concerns (Fagan, 2013b, 2015, 2016a). Here it is helpful to contrast the individuation conditions for *cells* with those of *stem cells*. The former are very straightforward: cells are individuated by an enclosing membrane. Cells are arguably the clearest example we have of a biological individual, in the sense of a living entity that can be counted, picked out from its environment, and distinguished from other such 'units'.[12] Cell individuality is topological: the cell is the membrane and that which it encloses.[13] The membrane boundary gives an unambiguous criterion for when we have one cell, two, or none. When cells reproduce by division, the process is finished when the membranes 'pinch off' to separate.

Stem cells, however, cannot be individuated by tracking membrane boundaries. As shown above, the very idea of a stem cell involves a cell lineage and differentiated states corresponding to that cell's 'future potential'. Indeed, the straightforward criterion for cell individuality actually conspires against individuating single stem cells. To show that a given cell is a stem cell, the defining capacities of self-renewal and differentiation need to be realized. But this cannot be done for a single cell.[14] To show that a cell is capable of self-renewal, an experimenter places it in a special environment that inhibits differentiation and observes whether that cell divides to make more stem cells. To show that a cell is capable of differentiating into a particular kind of differentiated cell, an experimenter places it in another environment which supports differentiation into that cell type ('along that pathway'). But it is not possible to perform both experiments on a single individual cell. In self-renewal, the starting cell is replaced by its progeny. When differentiation occurs, the cell is no longer a stem cell (and also, usually, has been replaced by its progeny). So stem cell researchers literally 'don't know what they've got 'til it's gone'.[15]

How then can stem cells be experimentally identified at all? Individuating stem cells is done at the population level. A population of identical stem cells (a 'clone') allows for a set of perfect replicate experiments. The idea is to test 'the same cell' across different environments, conducive to self-renewal and to a range of differentiation pathways. But this means that what we identify as a stem cell is also relative to the assumption that the population of cells we are testing is homogeneous, that they are 'the same cell' – not numerically, of course, but of the same type. In practice, this homogeneity assumption is unavoidable because it is not possible to experimentally individuate single stem cells. However, the criteria used to sort cell populations into homogeneous types ('subsets') are not guaranteed to be those shared by all and only the stem cells of interest. This is because we do not know the characters of stem cells in advance of experiments. There is no predetermined list of phenotypic characters that pick out all and only stem cells, of any variety; experimental researchers must work this out for themselves. This is done via repeated use of the experimental method using different combinations of cell surface markers, so as to gradually reduce mismatch with stem cell capacities. Stem cell researchers try to more and more finely specify the population of all and only stem cells of a particular type – to keep adding markers to get closer and closer to that pure population of, say, blood-forming stem cells in inbred mice (Fagan 2013a, Chapter 8). In this sense, experiments that identify stem cells are iterative.

Furthermore, claims about stem cell capacities that are well confirmed by experiment are highly subject to revision. This is because, as new cell traits are discovered and made accessible to measurement, the assumption that the cell population under test is homogeneous must be continually reassessed, and is often revised. So claims and hypotheses about stem cells are doubly relative: to details of the experimental method that specify values for variables in the abstract model, and to the homogeneity assumption for the population of cells tested. Both forms of relativity concern the experimental context. The first involves research materials and methods: the 'nuts and bolts' of biological practice. The second involves experimenters' epistemic situation: how much is known about the phenotype of the stem cells of interest, and how well these features match the criteria used to sort cells into populations in experimental tests of stem cell capacities.

It follows that stem cells, as biological individuals, are inextricably connected to particular experimental contexts. There is no 'absolute' or generally individuated stem cell type, as there is for, say, neurons or red blood cells.[16] To put the point particularly starkly, stem cells are *not cells* – that is, they are not individual 'units of life' that can be distinguished from their environment by a bounding membrane. Stem cells are instead the starting-points of lineages associated with particular experimental contexts. To conceive stem cells as the origins of lineages within the body of a multicellular organism involves a further move away from their experimental validation.

Lineage tree models

Alongside their experimental relativity, stem cell concepts and hypotheses involve substantive assumptions about biological development. At the cellular level, these assumptions can be made more explicit by further analysing variable D: a developmental process that orders cell states with respect to differentiation. Lineages are modelled using 'tree diagrams' that track relations between generations of reproducing entities.[17] Variable D can be conceptualized as a space of possible lineage tree models. Each model within this space has a particular number and arrangement of developmental stages, branch-points, and termini (see above). The number of stages distinguished is the 'depth' of the cell hierarchy for a particular lineage. The simplest cases, like asymmetric cell division, distinguish only two stages: a stem and a more differentiated state (Figure 8.2). Other developmental processes are more elaborate. The number of branch-points indicates the range of different pathways within a cell lineage. The number of termini of these pathways is a measure of the initiating stem cell's developmental potential.

The way branch-points are distributed among the stages leading to termini corresponds to a particular tree topology.[18] Some varieties of stem cell, such as those for skin and hair, initiate developmental processes with multiple stages but few branch-points. In contrast, stem cells of the blood and immune system form elaborate branching hierarchies. So we can characterize the topology of the cell developmental process D (the arrangement of cell states $s_1,...,s_n$) in terms of these sub-variables for a given stem cell: d, p, a. Any given process of development produces a cell lineage tree with a specific topology, representing developmental and reproductive relations between cells. Different stem cell concepts correspond to models that differentially constrain the space of possible topologies for cell lineage trees.

This framework provides a general way to systematically compare different varieties of stem cell, and to make explicit the assumptions about cell development that constrain interpretation of experimental data.

Multi-Level development

Stem cell concepts also involve substantive assumptions about development at organismal and molecular levels – more precisely, assumptions about cell-organism and cellular-molecular relations in development. Here I discuss the latter only. A cell state, recall, is a pattern of gene expression and molecular interactions that determines a cell's structural and functional characteristics. So, it is a multi-level concept. The concerns raised about stem cell individuation above require no assumptions about the molecular basis for cell identity. In much of biological practice, a cell's state is (or determines) that cell's identity: a molecular network of specific interactions in virtue of which a single cell is *the kind of cell it is*, whether that is a stem cell of any variety, or a neuron, fibroblast, etc. But, obviously, the process of development involves changes of state – that is, cell identity is in flux.[19]

We do not (yet) have a molecular account of 'stemness', so we cannot identify or individuate stem cells in terms of a particular molecular network. But there is a general way of representing the relation between cell and molecular levels: as a landscape (Waddington, 1957; Fagan, 2012). Waddington famously depicted the developmental potential of a tissue as:

> a more or less flat, or rather undulating surface, which is tilted so that points representing later states are lower than those representing earlier ones … Then if something, such as a ball, were placed on the surface, it would run down toward some final end state at the bottom edge.
>
> (1957, 29)

A rolling ball's path down the incline corresponds to the development of some part of an organism from an early undifferentiated state to a mature differentiated state. The bottom edge describes a series of dips representing alternative mature states, while the top edge describes a curve with a single minimum representing the undifferentiated start of development. The undulations of the landscape carve it into valleys that have the form of branching tracks, fanning out as they descend. These branching valleys connect the initial state with multiple discrete end-states. So the landscape represents developmental potential gradually restricted over time, partitioned into diverging channels (Figure 8.6). The landscape is a lineage tree model, with the developmental pathways 'recessed' into the terrain.

The inter-level relation of molecules and cells is visualized by the landscape's two sides: a top-side view of branching pathways leading to stable developmental states (lineage trees, embedded in a 'potential landscape'), and an underside of genes and their interacting products. This represents

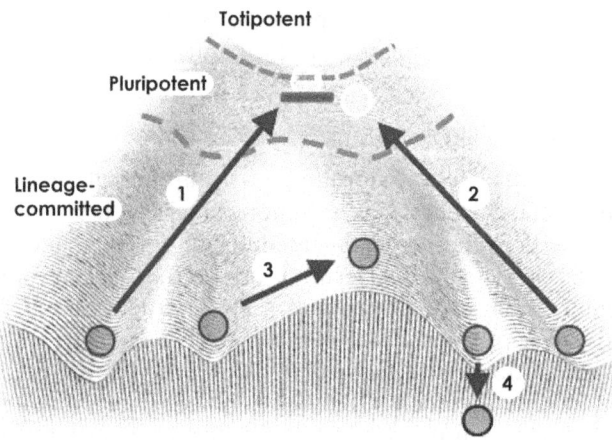

Figure 8.6 Updated version of Waddington's landscape, showing developmental
      pathways associated with reprogramming experiments.
Source: from Yamanaka 2009, 50. Reprinted by permission from Nature Publishing Group.

the idea that cell phenotype is determined by a particular pattern of gene
expression, which is conceptualized as that state of a molecular network:
a combination of molecules in the cell that interact in particular ways to give
the cell its distinctive features and behaviour. An explanatory model of cell
development would account for the dynamics of these states. A cell starts
in the stem state and from there enters one or another pathway, which leads
to a differentiated terminus. The landscape model thus raises a fundamental question for developmental biology: what controls the dynamics of cell
states? That is, what *causes* the succession of cell states to follow the specific
pathways emanating from stem cells? Is it a predetermined genetic program
encoded in DNA sequences? Or is it an orchestrated set of environmental
stimuli, from cells' physical environment and from one another? Is development driven by molecular interactions within a cell, among cells, from the
physical and biochemical environment, or a combination of these?

Waddington's landscape dodges the question, putting the cell lineage tree
on a 'tilt', which amounts to assuming a force of development as ineluc-
table as gravity. But we have learned, in part from stem cell biology, that
cell development is in fact reversible, and much more plastic and malleable
than traditional views of the process would suggest. Cell reprogramming is
a method of producing stem cells (induced pluripotent stem cells) from more
differentiated cells, effectively reversing normal development (Takahashi
and Yamanaka, 2006). This experimental discovery contradicts any strict
principle of development as irreversible and internally directed, undermining the notion of a fixed 'program for development' (Brandt, 2012). The stem
cell concept and its experimental applications present us with fundamental
questions about the nature of development.

Conclusion

I have presented a minimal abstract model based on the general scientific definition of 'stem cell', according to which a stem cell is the origin (or stem) of a cell lineage L, generated by n cell divisions and organized by comparison of characters C, which can be mapped onto a developmental process D of ordered cell states $s_1,...,s_n$. Different combinations of values of these variables correspond to different varieties of stem cell (adult, embryonic, pluripotent, induced, neural, muscle, skin, blood, etc.), which are currently targets of research. So the abstract model offers an inclusive (although hardly exhaustive) overview of the field of stem cell research, indicating its major strands and areas of evidential concern. The model also has implications that bear on current debates in philosophy of science and philosophy of biology, a few of which are briefly discussed above. In particular, the minimal stem cell model provides a framework for making explicit substantive assumptions about the process of development at cellular, organismal, and molecular levels.

The nature of developmental processes is also at the heart of ethical debates about stem cell research. The main focus of such debate is construction and use of hESC lines, which disrupts the development of a human embryo *in vitro*. Ethical viewpoints about the permissibility of this procedure (for human embryos) are premised on ideas about developmental potential, and whether the causes of development are localized within the developing entity or imposed from without. 'Inter-level' relations between stem cells, their organismal sources, underlying molecular circuitry, and organismal products are at the core of these ongoing debates. Any satisfying ethical position regarding stem cell research should directly engage these ideas. The abstract modelling framework presented in this chapter can, I hope, facilitate such engagement.

Acknowledgements

Thanks to Christine Hauskeller, Arne Manzeschke, and Anja Pichl for the invitation to contribute a chapter to this timely volume, and for their comments on an earlier draft. Ideas related to this paper were discussed with Hanne Andersen, Carol Cleland, Carl Craver, Carrie Figdor, Richard Grandy, Sara Green, Matt Haber, Lucie Laplane, Sean Morrison, Anne Peterson, and Joe Rouse; this paper has greatly benefited from their thoughtful comments. Related ideas were also discussed in presentations to the Committee on the History and Philosophy of Science (University of Colorado at Boulder), and the Science and Society Program (Wesleyan University), the Department of Philosophy (Wesleyan University), and at Worlds Apart? Aristotelian and Contemporary Engagement in Metaphysics and the Philosophy of Biology (University of Utah). Many thanks to participants at these events for valuable comments and discussion. Funding was provided by the Humanities Research Center at Rice University, the Mosle Research Foundation, the University of Utah College of Humanities, and a generous donation from the family of Sterling M. McMurrin.

Notes

1 For historical analysis of the stem cell concept and associated research, see Dröscher (2012) and Maehle (2011).
2 I use the term 'model' in a sense close to Giere (2004) and Weisberg (2013): an abstract or concrete object used to gain knowledge of a target system that it represents, in respects and degrees in accordance with model-users' purposes. The purpose of my abstract stem cell model, as stated in the main text, is threefold.
3 Melton goes on to expand on the general definition:

> a more complete functional definition of a stem cell includes a description of its replication capacity and potency … a working definition of a stem cell line is a clonal, self-renewing cell population that is multipotent and thus can generate several differentiated cell types.
>
> (2013, 7–9)

This more elaborate definition retains self-renewal and differentiation as the defining capacities of stem cells.
4 See also the *Cell Therapy and Regenerative Medicine Glossary* (2012), *National Institutes of Health Stem Cell Glossary* (2016), and further references in Fagan (2013a).
5 See Schwann (1847) for an early statement of cell theory. Theses (i)–(iii) are updated to reflect our current scientific context.
6 Gamete cells of dioecious organisms can also fuse to form a new cell: a zygote. But stem cell phenomena are limited to divisions that occur within one generation of organisms. So only mitotic divisions are discussed here.
7 Cell division events are classified as 'symmetric' if two offspring cells resemble each other, 'asymmetric' if they differ. Self-renewing divisions can be symmetric (both offspring resemble the parent) or asymmetric (one offspring resembles the parent).
8 Different assignments of values to characters C might correspond to different cell states, but it is important not to conflate the processes of cell division and cell state transition.
9 This abstract model is a modification of that presented in Fagan (2013a, 2013b, 2016b). The main contrast is the explication of differentiation. The earlier treatment stipulated that each mature cell type is distinguished by a specialized set of character-values M, which includes morphological, functional, and molecular characters. Cells in lineage L *differentiate* over some time interval t_1 to t_2 just in case cell character values for M at t_2 are more similar to those of mature cells than at t_1. The present account is a further generalization of that earlier treatment.
10 I return to this point below.
11 The value of variable D is partly so specified (Table 8.1), although the sub-variables discussed above are substantive assumptions about the process of development, rather than details of an experimental procedure. I discuss these assumptions below.
12 See Guay and Pradeu (2016) for a range of views about individuality in biology, physics, and analytic metaphysics.
13 Of course, signals from the environment pass through the cell membrane all the time, and vice versa – the membrane is not a barrier.
14 More detailed versions of this argument appear in Fagan (2013b, 2015).
15 With apologies to Joni Mitchell.
16 Also, the field of stem cell research is a patchwork of experiments, each supporting different specific hypotheses relative to the method used. In this sense, the field is inherently disunified.
17 See Dröscher (2014) for a historical study of these diagrams.

18 Inference from a set of termini and their traits to the correct tree structure is a core project of phylogenetic systematics. Tree diagrams have long been a unifying framework for development and evolution (see below).

19 The organism-level, which would tie all these changes into a single persisting individual, is conspicuously absent in much of stem cell biology. So development is seen as disarticulated and disunified.

References

Brandt, C. (2012). Stem cells, reversibility, and reprogramming. In: R. Mazzolini and H. J. Rheinberger, eds., *Differing Routes to Stem Cell Research: Germany and Italy*. Bologna: Società editrice il Mulino, and Berlin: Duncker & Humblot, pp. 55–91.

Cell Therapy and Regenerative Medicine Glossary (2012). Stem cell. *Regenerative Medicine*, 7, S1–S124.

Clermont, Y. and Leblond, C. P. (1959). Differentiation and renewal of spermatogonia in the monkey, Macacus rhesus. *American Journal of Anatomy*, 104, pp. 237–273.

Dröscher, A. (2012). Where does stem cell research stem from? In: R. Mazzolini and H. J. Rheinberger, eds., *Differing Routes to Stem Cell Research: Germany and Italy*. Bologna: Società editrice il Mulino, and Berlin: Duncker & Humblot, pp. 19–54.

Dröscher, A. (2014). Images of cell trees, cell lines, and cell fates: the legacy of Ernst Haeckel and August Weismann in stem cell research. *History and Philosophy of the Life Sciences*, 36, pp. 157–186.

European Stem Cell Network (2016). *Stem Cell Glossary*, [online]. Available at: www.eurostemcell.org/stem-cell-glossary#letters [Accessed 13 Mar. 2016].

Fagan, M. B. (2012). Waddington redux: models and explanation in stem cell and systems biology. *Biology and Philosophy*, 27, pp. 179–213.

Fagan, M. B. (2013a). *Philosophy of Stem Cell Biology*. London: Palgrave Macmillan.

Fagan, M. B. (2013b). The stem cell uncertainty principle. *Philosophy of Science*, 80, pp. 945–957.

Fagan, M. B. (2015). Crucial stem cell experiments? Stem cells, uncertainty, and single-cell experiments. *Theoria*, 30, pp. 183–205. (Special Section: Philosophy of Experiment)

Fagan, M. B. (2016a). Cell and body: individuals in stem cell biology. In: A. Guay and T. Pradeu, eds., *Individuals Across the Sciences*. Oxford: Oxford University Press, pp. 122–143.

Fagan, M. B. (2016b). Generative models: human embryonic stem cells and multiple modeling relations. *Studies in History and Philosophy of Science Part A*, 56, pp. 122–134.

Giere, R. (2004). How models are used to represent reality. *Philosophy of Science*, 71, pp. 742–752.

Guay, A. and Pradeu, T., eds., (2016). *Individuals Across the Sciences*. New York: Oxford University Press.

Hopwood, N. (2005). Visual standards and disciplinary change: normal plates, tables and stages in embryology. *History of Science*, 43, pp. 239–303.

International Society for Stem Cell Research (2016). *Stem Cell Glossary*, [online]. Available at: https://www.closerlookatstemcells.org/patient-resources/resources-from-isscr/stem-cell-glossary/ [Accessed 6 June 2019].

Maehle, A. H. (2011). Ambiguous cells. *Notes and Records of the Royal Society*, 65, pp. 359–378.

Melton, D. (2013). Stemness: definitions, criteria, and standards. In: R. Lanza and A. Atala, eds., *Essentials of Stem Biology*, 3rd ed. San Diego: Academic Press, pp. 7–17.

National Institutes of Health, U. S. Department of Health and Human Services. Glossary. Stem Cell Information, (2016), [online]. Available at: http://stemcells.nih.gov/glossary/Pages/Default.aspx [Accessed 13 Mar. 2016].

Potten, C. S. and Lajtha, L. G. (1982). Stem cells versus stem lines. *Annals of the New York Academy of Sciences*, 397, pp. 49–61.

Ramalho-Santos, M. and Willenbring, H. (2007). On the origin of the term 'stem cell'. *Cell Stem Cell*, 1, pp. 35–38.

Schwann, T. H. (1847). *Microscopical Researches into the Accordance in the Structure and Growth of Animals and Plants*. London: The Sydenham Society.

Takahashi, S. and Yamanaka, S. (2006). Induction of pluripotent stem cells from mouse embryonic and adult fibroblast cultures by defined factors. *Cell*, 126, pp. 663–676.

Thomson, J., Itskovitz-Eldor, J., Shapiro, S., Waknitz, M., Swiergel, J., Marshall, V. and Jones, J. (1998). Embryonic stem cell lines derived from human blastocysts. *Science*, 282, pp. 1145–1147.

Waddington, C. H. (1957). *The Strategy of the Genes*. London: Taylor & Francis.

Weisberg, M. (2013). *Simulation and Similarity: Using Models to Understand the World*. New York: Oxford University Press.

Yamanaka, S. (2009). Elite and stochastic models for induced pluripotent stem cell generation. *Nature*, 460, pp. 49–52.

9 What keeps an outdated stem cell concept alive? A search for traces in science and society

Anja Pichl

Abstract

This chapter argues that there is a common understanding of stem cells whose influence on both science and society is strong and problematic: the image of stem cells as clearly identifiable entities with certain intrinsic properties that can be used for therapeutic purposes. In the first part I shortly sketch how it prevails in society and reinforces certain problematic dynamics therein such as stem cell hype. In the second part, I shall investigate its role in science, drawing on debates about stem cell concepts among philosophers of biology and stem cell biologists. This analysis reveals conceptual and epistemic problems in studying and clinically applying stem cells on the basis of which their common understanding can be criticized as being untenable. That it nonetheless persists can be explained by two constituents of the field of stem cell research: clinical goals and methodological reductionism. I conclude that both of these (1) have a somewhat misleading influence on the understanding of stem cells and the choice of concepts and methods in studying them, (2) set too narrow limits on the scope and depth of stem cell research, and (3) contribute to unreasonable expectations which in the end put patients and science's reputation at risk.

Introduction

What are stem cells? This seemingly simple question is a matter of long-standing controversy among scientists and philosophers of science. Upon the answer to it, that is, upon the explicit or implicit understanding of stem cells, depend different ways of studying them and conceiving of the developmental and regenerative processes they are involved in. Different stem cell concepts also pave different pathways to the clinic, and to a differing degree warrant or question therapeutic hopes and expectations. That the field has still not delivered on its therapeutic promises has also been related by stem cell scientists themselves to problems of understanding stem cells (Dimmeler et al., 2014). Therefore it is not just an abstract philosophical problem that 'stem cells are widely discussed yet poorly understood' (Fagan, 2013b, 945). This lack of understanding and ambiguity gives more than usual power to certain preconceptions and images that (in)form certain stem cell concepts.

Under the guise of a purely scientific concept, the multiple influences between the closely entangled spheres of science and society that contribute to and build on the common understanding of stem cells remain invisible. Their critical scrutiny is a desideratum this chapter aims to address in recourse to insights provided by philosophy and social studies of science.

Stem cells are usually conceived as clearly identifiable entities with certain intrinsic regenerative properties (referred to as the 'entity-view' in the following) that will be brought under researchers' and clinicians' control for therapeutic purposes. This common understanding prevails in the wider public, among patients and policy-makers, yet also in science itself. It will be sketched in the first part of this chapter, together with some societal problems and dynamics that have had an influence on or were reinforced by it. In the second part, its roots in, and consequences on, stem cell research will be presented – and in particular its tenability examined. Philosophers of science and scientists discussing stem cell concepts revealed epistemological problems and evidential constraints that culminated in the discovery that 'stem cells are not cells' (cf. Fagan, in this volume) – the opposite of the entity-view. Having shown with recourse to these debates that the common understanding of stem cells relies on rather shaky epistemic grounds, the question arises as to how it can still be so prominent. In the third part of this chapter I shall trace its continued influence back to two constituents of the research field: clinical goals and basic reductionist commitments. I shall tentatively conclude that, notwithstanding a great many valuable research results they made possible, both clinical goals and reductionist commitments (1) have exerted a misleading and narrowing influence on the understanding and study of stem cells, that they (2) might therefore even hinder the development of some clinical applications by foreclosing the search for alternative ways of reaching them, and (3) that the entity-view of stem cells appears to be epistemically not well supported and to reinforce problematic dynamics in both science and society. How to understand and handle limits of reductionism in stem cell research and how to properly build societal goals into science are questions that have to be left open for further research.

This chapter stands in the wider context of the question as to how philosophical analyses of scientific research can be socially relevant. It largely follows Fehr's and Plaisance's proposal that by 'clarifying key concepts in socially relevant science, identifying questionable methodological assumptions, pointing to epistemic failures and suggesting improvements, or determining epistemic reasons for why potentially useful scientific knowledge is not being given uptake' (Fehr and Plaisance, 2010, 308), philosophy of science might produce some potentially enlightening and societally significant results by its own means, especially if the close entanglement of science and society is taken into account in the conceptual analysis as has been done by Hauskeller and Weber in their study of the role of ethical concerns, therapeutic expectations, and scientific uncertainty in how scientists frame and present different types of pluripotent stem cells (Hauskeller and Weber,

2011). Highlighting some still unresolved conceptual and methodological issues and their relation to societal dynamics and expectations will obviously not enable reliable predictions of concrete future outcomes or courses of research. But it might help to overcome some poorly justified assumptions about stem cells and their capacities that inform current scientific practice as well as public and ethical debate, and that contribute to certain societal problems. I shall proceed by focussing on the tensions between common notions of stem cells in science and society on the one hand, and philosophical accounts and scientific insights on the other hand, touching on the wider significance of these tensions and ambiguities.

The prevalent societal understanding of stem cells

Stem cell research has long been, and still is to some extent, represented as *the* great therapeutic hope for science and society that will provide a 'third leg' to medicine (Knoepfler, 2013): regenerative cell therapies in addition to conventional drugs and surgery. The promise of stem cell therapies against a broad range of diseases (cf. Sontag and Zenke, and Tanner et al., in this volume) has been made and nurtured by scientists and policy-makers, ethicists, and funding agencies alike. I shall draw on Burns' (2009) analysis of the stem cell debate in the early 2000s for clarifying the common understanding of stem cells. Burns (2009) coins the term 'stem cell superhero' in order to characterize stem cell promises which differ significantly from the traditional magic bullet model (Stegenga, 2018) with regard to a certain transformability and agency ascribed to stem cells. Burns exemplifies the diagnosed implicit use of the superhero metaphor at work in media, science communication, and public deliberation about stem cell research by four essential traits:

> 1. Stem cells have a seemingly "magical" capacity to heal virtually any disease, drawing on unprecedented powers and the unique skills of gifted persons; 2. There is no other way to fight the evil we face (this is the only hope for relieving immense suffering); 3. All that stands in the way of achieving unprecedented therapeutic benefits is a willingness to summon the superhero (by letting the researchers do their work); 4. Those who stand in the way of the superhero are unwittingly aiding the cause of evil (in other words, opponents of stem cell research are unintentionally causing millions to suffer needlessly). Each trait has a special contribution to make: (1) indicates the magnitude of the benefit, (2) adds urgency, necessity, and inevitability, (3) signals the path to take and the proximity of the cure, and (4) assigns blame and exerts pressure.
> (Burns, 2009, 429)

Anyone familiar with the public discourse on stem cells in the early 2000s will easily recognize these features. The 'superhero' image of stem cells certainly played a major role in raising funding and public support for this field of research. The enormous societal resonance to its therapeutic visions – or rather the 'convergence of expectations' (Burns, 2009, 435) of both scientists and the public – contributed to its worldwide institutional establishment in a very short period of time. Within the last twenty years, significant progress towards a better understanding and handling of human stem cells has been made, including steps towards clinical application and a range of scientific discoveries such as the Nobel Prize-winning development of induced pluripotent stem cells (cf. Chapter 5, in this volume). Thus the 'superhero' image of stem cells served well in raising public support and allowing a great community of researchers to concentrate their efforts on this emerging field. But it has a rather problematic – and, as the next section argues, epistemically untenable – conceptual core: as Burns' first trait and the examples in his paper show, stem cells are being attributed 'a "magical" capacity to heal virtually any disease' (Burns, 2009, 429). In other words, stem cells are conceived of as clearly identifiable cell types with intrinsic capacities usable for a broad range of therapeutic purposes. This understanding is widely shared among scientists, patients, and the wider public. It appears to be also deeply ingrained in the standard scientific definition, which functionally characterizes stem cells by their potential of differentiation and self-renewal. The stem cell thus conceived as being 'defined by what it gives rise to rather than its observable traits' (cf. Chapter 8, in this volume, p. 112) has a promissory character in its conceptual core mirroring its role as a 'superhero' in societal perception.

The entity-view of stem cells can also be identified at the base of the endowment of bio-value to stem cells and of ethical and regulatory debates that put much moral and legal weight on the allegedly clear distinction between different cell types, e.g. totipotent and pluripotent cells. It has been massively advertised and fulfilled further functions apart from providing scientists with resources: stem cell promises, closely tied to supposedly cell-intrinsic capacities, were effective in calming down ethical qualms regarding the use of human embryos for research – partly by exerting moral pressure on opponents as Burns' fourth trait illustrates – and in overcoming regulatory hurdles to the study of human embryonic stem cells (hESC). Yet the argumentative use of expected future outcomes of research not only remained rather speculative: ethicists often simply repeated scientists' views and singled out one problematic aspect – the moral status of the human embryo – leaving the rest of the therapeutic promises, scientific concepts, assumptions, aims, and values of the research field largely unquestioned (cf. Hauskeller, 2002 for an early criticism). They also contributed to stem cell hype: by making a moral argument of the therapeutic potential of embryonic or adult stem cells respectively (cf. criticism by Lysaght and Campbell, 2013; Towns and Jones, 2004), both advocates and opponents of embryonic stem cell research treated envisaged stem cell therapies prematurely as a scientific fact rather than as a promise, and encouraged 'understandings in the public discourse that "stem cells" not only hold great promise but that the miracle cures they might produce are just around the corner' (Lysaght and Campbell, 2013, 256). Further factors such as myths of the power of nature and of high technology as a means for scientists to unlock stem cells' natural potential (Burns, 2009; cf. Chapter 2, in this volume) also had a considerable impact on building extensive promises. The consequent high expectations played a crucial role in the establishment of a market of stem cell treatments without proven safety and efficacy which have been offered to patients all over the world (cf. Chapter 2, in this volume; and Turner and Knoepfler, 2016 for the dimension in the USA alone).

In recent years, researchers, ethicists, and policy-makers have become increasingly concerned about the flourishing market of unproven stem cell treatments they deem to be a danger for patients' health and science's reputation. They have criticized stem cell science communication strategies as fuelling hype, and officially modulated them: the 2016 'Guidelines for stem cell science and clinical translation' from the International Society for Stem Cell Research (ISSCR) put a new (in comparison to the version from 2008, cf. International Society for Stem Cell Research 2008 and 2016) and strong emphasis on more cautionary discursive strategies and framed them as a part of scientific integrity. The aim was to close the 'gap between public expectations and the actual state of stem cell science and clinical development' (Caulfield et al., 2016). But the genie seems to be out of the bottle: the constant rise of the number of private clinics selling stem cell treatments without proven safety and efficacy worldwide indicates that their thousands of clients' high expectations are still unshaken, notwithstanding the official warnings and accessible patient information such as the *ISSCR Patient Handbook* (2008). Equally high confidence in stem cells' and researchers' capacities is visible in regulatory efforts to accelerate the translation 'from bench to bedside' by lowering regulatory thresholds for entering clinical trials and market admission (cf. Chapter 12, in this volume; cf. *Nature* Editorial, 7 July 2016). These regulatory changes attempt to advance the translation of stem cell research into medical treatments by lowering or even abolishing established standards of evidence of safety and efficacy such as randomized controlled trials (RCTs) for introducing new drugs and therapies. Notwithstanding widely discussed epistemic and practical problems of RCTs (Stegenga, 2018): simply sidestepping established forms of providing evidence without substituting them with some good alternative seems problematic and to rely on rather strong beliefs about stem cells' capacities and the readiness of regenerative medicine. Recent efforts to confine stem cell hope and hype within reasonable limits by way of more balanced science communication are certainly helpful, but also part of the problem they attempt to solve because they do not include a self-critical reflection of stem cell science and its conceptual, methodological, and ideological basis. The next section examines what philosophy of science debates on stem cell concepts and stemness can contribute to this matter.

Stem cell concepts in science and philosophy of science

Stem cells: state or entity?

The prevalent societal understanding of stem cells has its roots in, and many resemblances to, the 'classical' scientific stem cell concept: '[w]ithin the classical view of stem cells, the property of stemness is considered an intrinsic property – that is, stem cells are a natural kind or type that are inherently endowed with the ability to self-renew and to differentiate' (Laplane, 2016, 180). This concept of stem cells comes closest to the sketched societal entity-view and is still 'widely accepted throughout biology' according to Laplane (2016, 181), albeit contested:

> there have been calls from stem cell biologists to recognize stemness as an extrinsic property that non-stem cells can acquire, arguing that the concept of stem cell should be understood as referring to a "state" (Zipori) or "function" (Blau) rather than to an "entity". This move to recognize stemness as an extrinsic property, conferred upon stem cells rather than inherent to them, pushes stem cells away from being a determined cell type and toward stem cells being a cell state, characterized by reversibility and transiency.
>
> (Laplane, 2016, 181)

The long-standing state-versus-entity debate was based on the distinction of stemness as a cell-intrinsic or -extrinsic property. Attempting to advance and clarify the underlying ontological presuppositions, Laplane distinguishes four different ontological concepts of stemness taking into account the role of the niche and of the dedifferentiation potential: (1) *categorical* ('a property that is essential and intrinsic to the cells belonging to the stem cell natural kind') (Laplane, 2016, 155), (2) *dispositional* (an intrinsic property depending on external stimuli), (3) *relational* (an external property arising from the microenvironment), or (4) a *systemic* property (an external property controlled at the system level) (Laplane, 2016). She investigates the prospect of success of the envisaged therapeutic strategies if the different stem cell concepts were adequately representing cancer stem cells; the central question being whether the right therapeutic target is the cancer stem cell, its microenvironment, or both. Also for envisaged stem cell regenerative therapies, the questions whether stem cells are entities or states different cells can enter and leave, and what role the niche plays regarding stem cell identity and functioning, are crucial for assessing the likeliness that transplanted cells will work in the bodily environment in the supposed way, whether they will lose their regenerative capacities in the new micro-environment, give rise to cancer, or have other side-effects (cf. Laplane, 2016, 188–9). The relevance of the understanding of stem cells for scientific practice and therapeutic application has thus been convincingly shown, but the question remains whether the level of stem cell ontology in combination with sharp classificatory definitions is the most promising road to clarify how to adequately conceptualize stem cells. Laplane's agnosticism regarding the question as to which of her four stem cell concepts would best fit the phenomenon gives rise to doubts. If for scientists the adoption of one concept is methodologically necessary, Laplane suggests choosing the most exigent one, i.e. the systemic conception – according to which stemness 'is induced by some heterogeneous intrinsic and extrinsic mechanisms at the population level' (Laplane, 2016, 178) – leading, according to her, to most open research (Laplane, 2016, 187). This may be so, but the idea that all four concepts were equally plausible and that researchers interested in basic phenomena of stem cell identity and functioning could possibly and fruitfully decide themselves for one fixed concept, and arbitrarily shape their research according to it is not convincing. The epistemic problems discussed in the next section and the widely acknowledged 'reconceptualization of the cell as a "state" rather than a fixed "entity"' (Kraft and Rubin, 2016, 519) in the course of research into cellular differentiation and plasticity especially question the tenability of the categorical stem cell concept.

Furthermore, as Rheinberger (2006, 221ff.) argued, scientific terminology is necessarily vague and context-dependent, thus allowing scientists new insights into epistemic objects. Rather than attempting to sharply define scientific concepts with the hope to 'introduce clarity and precision into research' (Laplane, 2016, 9), Rheinberger suggests studying the experimental systems and epistemic practices which incorporate vague and productive scientific terminology. With her proposed 'binominal definition', Laplane (2016, 112ff.) tries to reconcile this need for an open stem cell concept with fuzzy boundaries with the apparent indispensability of clear and precise definitions required for therapeutic application. Her combination of a very broad term of stem cells as 'cells from which tissues develop and are maintained' (Laplane, 2016, 113), with a multitude of subtypes and sub-subtypes of specific stem cell concepts is certainly very useful, e.g. for ordering the stem cell types discovered and characterized by researchers. But it seems too descriptive to answer fundamental questions such as what stem cells are and how they relate to their environment. Neither the definition nor the ontological categories provide a criterion for distinguishing whether stemness can be understood as an intrinsic property of stem cells or not. Furthermore, it neglects that the results and methods of science, and the aims and horizon of research, strongly depend upon certain preliminary understandings of the object. Therefore, some caution is due in making inferences about stem cell ontology based on the latest research methods and results.

Elusiveness and evidential constraints

Stem cells have been shown to be quite 'elusive' (Robert, 2004, 1007; Zipori, 2004, 877): *they lack specific properties or features shared by all and only stem cells*, whereas other cell types in the body such as neurons or blood cells are defined and easily identified by certain observable properties (cf. Chapter 8, in this volume, p. 124). The many different types of stem cells within different tissues of the body at changing developmental stages and within different species have no unifying feature. Consequently, '[n]o stem cell is a model for stem cells as such' Robert (2004, 1008) emphasized, explaining arising problems of extrapolating results from stem cell experiments to stem cells in other contexts:

> 1. Where our models of stem cells are sourced at a particular developmental stage (embryonic, fetal, or 'adult'), we must be cautious in making generalizations to other developmental stages. 2. Where our stem cells are derived from a particular developmental system (e.g., blood, pancreas, brain, skin), we must be cautious in making generalizations to other developmental systems. 3. The existence of differences in stem cell behaviour and morphology between, say, mouse and human, indicates that we must be cautious in making generalizations across species boundaries. 4. Highly derived cell lines may be importantly different from their 'wild-type' counterparts, such that we must be cautious in making generalizations even within species. 5. It may well matter whether our studies have been conducted in vitro or in vivo, and what culture media have been used in generating and maintaining the cell lines in question.
>
> (Robert, 2004, 1009)

Results from *in vitro* studies cannot simply be extrapolated to the living organism or other experimental settings, and it cannot be presupposed how the multitude of stem cells function in different locations, and how they become activated or surpassed during diverse stages of the development of an organism. Consequently, stem cell researcher Flake proposed the following maxim for interpreting experimental results:

> we must confine the interpretation of an experiment to what has actually been demonstrated. To extrapolate conclusions about normal biology or events in specific biological systems from in vitro assays is inherently illogical and has a very high probability of simply being wrong.
>
> (Flake, 2004, 62)

Accordingly, many stem cell researchers emphasize the desideratum to study stem cells in their context, i.e. 'how stemness arises not only as a cell-autonomous property but rather, as a property emerging from the behaviour of cell population' (Alvarado and Yamanaka, 2014, 112–3). There have already been obtained astonishing results by *in vivo* imaging of adult neurogenesis which call for reconsidering the notion of stem cells as starting points of a unidirectional cell lineage (Goetz, 2018) – one of the basic tenets of stem cell research.

The outlined problems of inference seem not to be sufficiently taken into consideration by adherents of the traditional entity-view of stem cells. Notwithstanding the mentioned diversity and context-dependence of stem cells, researchers have sought to characterize and distinguish stem cells from other cells by searching for a specific gene expression profile supposedly implicated in self-renewal and differentiation (at least during these division events). This research was based on the two assumptions that 'stemness (self-renewal and differentiation) qualitatively distinguishes stem cells from non-stem cells' and that 'stemness is reducible to a set of molecular properties' (Laplane, 2016, 115). But the search for a general molecular signature of all and only stem cells by three research groups in the early 2000s failed, resulting in lists of 216–385 overexpressed genes of which only one gene that is not stem-cell specific was common to all three lists (Laplane, 2016, 119). Many researchers referred to methodological constraints of these studies to explain this negative outcome (e.g. Leychkis et al., 2009), whereas others questioned the existence of a common molecular stem cell signature and the reasonability of the search for it, cautioning that 'the investigator might be forcing the stem cell phenotype on the population being studied' (Zipori, 2004, 876). Considering the enormous diversity of stem cells in nature and cell cultures mentioned above, and the vagueness of this concept (cf. Tajbakhsh, 2009), there is most likely a continuum between stem cells and non-stem cells with no rigorous boundaries, and many different molecular mechanisms might be employed by different stem cell types as Zipori argues (2004, 875).

Without certain measurable properties and no stem cell signature at hand, scientists must experimentally 'realize' the developmental potential of stem cell candidates in order to identify and characterize stem cells with certainty. Despite the variety of experimental techniques to do so, some basic steps are common to all stem cell experiments. These steps include the removal of cells from tissue and the measurement of characters in different environments conducive to self-renewal or differentiation (cf. Chapter 8, in this volume). The field of stem cell research is thus unified by common experimental practices rather than by a shared object, a shared theory of stem cells, or a theory of biological development, as Fagan argues. She highlights that these experimental steps face significant evidential constraints: self-renewal and differentiation potential can neither both nor separately be assessed for a single cell (cf. Chapter 8, in this volume). Consequently, 'you literally don't know what you've got 'til it's gone' Fagan (ibid.) notes when describing the 'stem cell uncertainty principle'. The strategy stem cell scientists employ in handling this limitation is to identify stem cells at the population level, using supposedly identical replicas of the same cell within a supposedly homogenous cell culture, more and more specified through iterative testing. But as no list of predetermined characters of stem cells exists, Fagan argues that the assumption that the researcher actually has a homogenous population of identical cells is provisional. Thus the identification of a stem cell is relative not only to the experimental context but also to the assumption that the population is homogenous, which might in practice be rather unlikely as some stem cell researchers observed: '[a]ll stem cell populations are inherently heterogeneous and function in the context of a network' (Flake, 2004, 62).

How closely stem cells' identity and functioning depend upon their context is further indicated by their 'developmental versatility', meaning their 'ability to produce different aspects of organization associated with different modes of development: normal, disordered, simple, cellular, organ-like, and embryo-like' (Fagan, 2017, 19). Which mode of development is realized in a particular stem cell-initiated process depends on the cell's and its descendants' environment, e.g. on 'the presence or absence of specific biochemical signals, spatial/geometric arrangement of cells, physical boundaries of the system, and physical factors such as oxygen level and pH' (Fagan, 2017, 19). The insights into the context-dependence and non-fixedness of stem cell identity and functioning are usually referred to as cellular 'plasticity'. In light of the experimental possibilities of changing cells' identity and potential as well as modes of development by changing their environment, the entity-view and its scientific equivalents seem rather outdated: 'it is cellular context, not intrinsic molecular specification, that establishes whether a cell type becomes a stem cell, a transit amplifying cell or simply goes extinct' (Lander, 2009, 70.5). A stem cell is not simply a biological individual with stable intrinsic properties and clear spatio-temporal boundaries, defined by its parts and features and delimited and distinguished from its environment by its membrane. A closer look at the standard, context-oblivious functional definition of stem cells as characterized by their potential for self-renewal and differentiation and its role in experiments leads Fagan to the conclusion that

> stem cells are *not cells* – that is, they are not individual 'units of life' that can be distinguished from their environment by a bounding membrane. Stem cells are instead the starting-points of lineages associated with particular experimental contexts. To conceive stem cells as the origins of lineages within the body of a multicellular organism involves a further move away from their experimental validation.
>
> (Fagan, in this volume, p. 124)

Problematic constituents of the field of stem cell research

In light of the discussed conceptual and scientific findings, the epistemic grounds for the entity-view prevailing in science and society seem rather unstable. So, why is it still so influential? In the following I shall show that its continued existence can be traced back to two important constituents of the field of stem cell research: methodological reductionism and clinical goals.

Reductionism

Since the beginning of modern biology, the limited adequacy of reductionist methods employed in scientifically studying biological organisms for understanding complex and context-dependent biological phenomena has been pointed out and criticized by many philosophers of science and life scientists (Brigandt and Love, 2017). Notwithstanding, experimental life science research today is deeply characterized by the reductionist methods of 'decomposition', 'internalism', and 'studying parts in isolation' (Kaiser, 2011). The discussed epistemic problems of identifying and characterizing stem cells have some roots in fundamental problems of methodological reductionism in biology. Stem cell research employs at least two forms of reduction. First, the complex biological processes of development and tissue repair are being reduced to and studied on the cell level mainly. Stem cells are conceived as 'active sources' (Fagan, 2013a, 3) of development and regeneration, and taken to be the right level and unit of analysis. Large parts of stem cell research are thus abstracting from the complex processes on the different biological levels from niche to whole organism or experimental system in which they are involved and which are constitutive for their identity, features, and functioning as discussed above. Stem cell researcher Lander (2009, 70.2) remarked critically that

> [s]o far, the main way in which researchers have sought to get a handle on stemness has been to try to reduce stem cell behaviour – a phenomenon operationally defined at the level of tissues and tissue reconstitution – to a set of necessary and sufficient, intrinsic cell-level properties.

Also Robert and colleagues criticized stem cell researchers' focus on the level of cells for being too narrow to provide an adequate understanding of the phenomenon because '[r]egulation and differentiation are properties of dynamic systems (cells and organisms) interacting with their respective environments' (Robert et al., 2006, 26).

Stem cell research is, secondly, characterized by decomposition of this already reduced whole into its constituent parts which are objects of the search for mechanistic explanations of their characteristics and interactions, usually in cell culture or model organisms, thus outside their original organismal context. Critics of reductionism in philosophy of biology point out that

> studying the parts outside of the systems provides only limited insights into the properties the parts exhibit in situ, how they are organized in

the system, and which interactions take place between the parts when the system displays the behaviour in question.

(Kaiser, 2011, 470)

The problems of extrapolating results from one system of stem cell research to others, or to the biological organism discussed above, are indicative of this limitation. Reductionist research strategies are usually taken to provide experimentally validated knowledge and the means to take over control of the processes studied, thus as the road to biomedical application. But:

> [t]he more complex and integrated a system is, the less adequate appear reductive explanations that refer exclusively to lower-level internal entities ..., and only to those properties and interactions between these entities that can be studied in other contexts than in situ.
>
> (Kaiser, 2011, 471)

Stem cells as starting points of cell lineages without clear spatio-temporal boundaries play a role in most fundamental processes of development and regeneration concerning all levels of the organism, so that reductionism seems to be especially problematic in stem cell research. Some researchers even postulate that 'we must stop applying reductionist principles to stem cell biology' (Flake, 2004, 62; see also Robert et al., 2006).

One could even go a step further and trace back the sketched reductionist commitments to an underlying substance ontology widely shared throughout Western science and society, following Dupré and Nicholson who propose an understanding of the living world as being actively maintained by processes, not made of substances with clear parts and specific properties, as more adequate (Dupré and Nicholson, 2018). Common substance ontology might bias stem cell scientists towards the entity view of stem cells. Even though the leap from the patchwork of scientific knowledge that highly depends upon context, epistemic interests, social goals, and further arbitrary factors seems just too big to subscribe to a specific ontology, it would certainly be worth the effort and heuristically useful to philosophically investigate the field of stem cell research from a process perspective. However, finding alternative, process-based or more holistic research methods as postulated by Robert and colleagues (2006), or Flake's urge to apply systems thinking to the study of stem cells (Flake, 2004, 59), face their own problems of viability and translation into concrete study programmes. There is a desideratum of studying how to better understand and handle reductionism and its limits in stem cell research.

Clinical goals

Clinical goals play a constitutive role in shaping the field of stem cell research (Fagan, 2013a, 224ff.) and thus they have arguably led to the popularity of the entity view of stem cells and further problematic assumptions that seem to serve these goals best: '[m]ost clinical aspirations of stem cell research presuppose that we can isolate and purify stem cells for medical use. Giving up this aim when research is still at an early stage would be drastic and rash' (Fagan, 2013b, 954). That is certainly true, but sticking too closely to it and to a possibly inadequate stem cell concept entails the risk of wrong preconceptions, especially in the light of the epistemic problems discussed above. It also distracts attention from other accounts of stemness and from issues such as the understanding of cell interaction and biological development. Neglecting these more fundamental questions or taking them in distorted ways for supposedly speeding up clinical translation has been argued to be detrimental to stem cell research and its clinical goals:

> it is the demand for translation itself that is driving the clinical studies while careful benchside study is often being ignored. ... We are in effect, translating from sadly incomplete benchside and bedside source languages, languages with unknown grammar, unknown syntax, and a few native speakers. In the end, translation may prove extremely difficult to achieve for stem cells in part because we do not know enough of the basic science.
>
> (Maienschein et al., 2008, 48)

The prevalence of clinical goals comes at the cost of a systematic study of the whole range of different cell types in different developmental stages and species, thus limiting the achievable insights into stem cells and development according to some philosophers and scientists (cf. Chapter 8, in this volume; Alvarado and Yamanaka, 2014; Robert, 2004). Most research concentrates on human beings as future beneficiaries and mice as standard model organisms.

A closer look at envisaged therapeutic strategies confirms that the transplantation of stem cells or their derivatives is still at the centre of therapeutic visions (cf. Chapter 5, in this volume; Dimmeler et al., 2014; Trounson and DeWitt, 2016). Knoepfler (2013) distinguishes five forms of how stem cells might be therapeutically applied: (1) cell therapy (transplantation of stem cells or their progeny), (2) immune modulation, (3) secretion of so-called 'trophic factors', (4) drug delivery, and (5) organ replacement. All these forms presuppose that stem cells can be clearly identified and that they have certain powers or intrinsic properties which they exert in different environments, e.g. stimulating immune responses and facilitating repair, or delivering drugs to injured tissue. The idea of taking over control and 'that if we can work out the genetic and molecular mechanisms, we can drive stem cells to work' (Roberts et al., 2006, 22) still characterizes many visions of stem cell therapies and shapes also basic research accordingly. But scientists are also exploring other ways of using stem cells or their plasticity (cf. Chapter 5, in this volume; Trounson and DeWitt, 2016), e.g. by stimulating or enhancing endogenous stem cells or their niches for tissue repair

(Dimmeler, 2014), using induced pluripotent stem cells as model systems for diseases (cf. Chapter 7, in this volume), *in vivo* reprogramming of cells to functional stem cells (cf. Chapter 6, in this volume; Heinrich et al., 2015), or targeting the niche or niche-cell interaction (Laplane, 2016). These novel approaches cannot be analysed within this chapter. However, the conceptual shift from entity view to an acknowledgement of stem cells' plasticity and context-dependence seems often only half done and rather strong views of stem cells or niche-cell interactions as more or less fixed, or at least as sufficiently fixable in order to be usable for prediction and control seem to prevail. This is visible in assumptions such as that niche-targeting or *in vivo* reprogramming will have the desired outcome: regenerating cells and tissues where necessary but not causing cancer or other side-effects.

The concept of cellular plasticity has also been associated with, and even strategically used to convey, therapeutic promises right from the start of its scientific use, as Kraft and Rubin (2016) have argued:

> Plasticity was serving as a means to articulate both new science and its perceived potential uses. On the one hand, it retained … its epistemological role within science as a means to conceptualize novel findings about cell differentiation and the changeability of cell identity. On the other hand, it was deployed by those wishing to develop stem cell therapies in ways that helped attract research funding and/or investment for commercial enterprises geared to this goal. … [P]lasticity was immediately and emphatically coupled to a vision of clinical utility.
>
> (Kraft and Rubin, 2016, 513)

Kraft and Rubin's article allows insight into a widespread attitude towards stem cell research in both science and society: with the possibility to cultivate and experimentally manipulate human stem cells, they seem to have been identified as therapeutic agents with little regard for theoretical and practical obstacles ahead. To put it in somewhat polemic terms: clinical utility was simply presupposed before thorough investigation of researchers' capacities to understand and control cellular plasticity. Confirmation of the suspected therapeutic power of stem cells seems to have been expected in the form of working therapies by many. In happy anticipation, patients and industry have pressed policy-makers (with often equally optimistic views) to lower regulatory thresholds for clinical trials and the commercial use of stem cell products. That cell identity is not fixed but changeable and differentiation reversible of course suggests therapeutic usability. But the therapeutic promises attached to the concept of plasticity seem to include rather selectively certain traits while disregarding others as well as the problem of how to foresee and govern cellular behaviour:

> [p]roponents of stem cell research are excited about stem cells precisely because stem cells are not fully determined, and because they respond

to their changing environments. But this, by definition, also means that we cannot easily predict how they will behave in various environments.

(Robert et al., 2006, 24)

Scientific and technological development may well overcome the present constraints of understanding and of handling stem cells; it can and should not be excluded that clinician-scientists develop stem cell cures without much understanding of the phenomenon. The case that a drug works without sound knowledge of the involved mechanisms is a case all too familiar to the medical community. But many envisaged stem cell cures rely on precise expectations of achievable knowledge and control, and on ideas of the functioning of stem cells that are highly questionable in the light of the outlined evidential constraints in identifying and characterizing stem cells and of the role of the cellular environment(s) for stem cells' developmental potential and versatility. Unlike drugs that are (sometimes) accepted to work in inexplicable ways and used after being checked for side-effects, stem cells as agents of a new regenerative medicine (Knoepfler, 2013) are expected to exert biological capacities which they alone are unlikely to have. Exploring the range of experimentally inducible changes to cellular differentiation processes and abilities to manipulate cell fate in experimental contexts is of course an important source of knowledge and might lead to therapies, but the road to stem cell cures and to an understanding of how the cellular environment and the whole biological organism regulate stem cell behaviour is still far off. The unresolved question as to how stem cells and stemness can be understood and properly studied is indicative of fundamental gaps in knowledge and of how to search for it in a specific field, all of which makes the supposed readiness of that scientific field to move into the clinic appear questionable. Of course stem cell research is nonetheless rapidly moving into the clinic with an ever-increasing number of ongoing clinical trials (cf. Chapter 5, in this volume; Trounson and DeWitt, 2016), but even some stem cell researchers have criticized this development for being too fast (Knoepfler, 2013). The question as to when to begin clinical trials and what might be the right criteria of evidence for deciding this question are highly contested among scientists and clinicians (Robert et al., 2006).

Conclusions

The guiding idea of this chapter was that the prevalent understanding of the stem cell as a specific cell type with certain intrinsic regenerative capacities is reinforcing problems in both science and society. I argued that this common understanding of stem cells is untenable in light of the discussed scientific insights and conceptual issues. Its continued influence, despite being epistemically problematic, can be traced back to its close relation to the clinical goals and basic reductionist commitments that are constitutive but also problematic for the field of stem cell research. To call the widespread preference of the entity-view of stem cells among scientists and the public alike, a case of wishful thinking as therapeutic visions hinge on it would probably be an overstatement. But the strong entanglement of science and society reflected in the understanding of stem cells raises at least the task of thinking about proper ways to build societal goals into science. Most scientists and philosophers of science embrace the role of clinical goals as being constitutive for stem cell research and uniting the field (Fagan, 2013a, 224ff.; but for criticism see Maienschein et al., 2008). Clinical goals are presented as good both for society and for science by assuring the field of stem cell research public acceptance, financial support, and societal relevance with the hope of delivering medical treatments. But certain perils and problems of this strong interlinking of science and society might benefit from further scrutiny. On the basis of this study of stem cell concepts and their roots in science and society, it seems at least plausible to argue that therapeutic goals, together with reductionist commitments, might (1) have a misleading influence on the understanding of stem cells and the choice of concepts and methods in studying them, (2) set too narrow limits on the scope and depth of stem cell research, thus possibly foreclosing some ways of understanding basic stem cell phenomena and alternative ways towards therapies, and (3) contribute to unreasonable expectations which put patients and science's reputation at risk. The pitfalls of a 'socially relevant science' and the tensions between the pursuit of knowledge and the pressure to generate stem cell treatments are in need of more thorough investigation than is possible here. If we take a closer look, stem cells, as the entities we imbue with therapeutic hopes and often also moral and economic values, kind of disappear. We are left with hardly understood organismal processes in need of further investigation. The limits of currently available knowledge about stem cells get frequently surpassed, e.g. when certain experimental practices and aims of research or societal expectations are based on ideas about stem cells, their capacities and functioning, that are biologically hardly plausible because they abstract from the cellular and organismal environment of stem cells or conceive of these relations as fixed and sideline unresolved epistemic problems. Also ethical and policy issues such as patient safety, regulatory, and funding decisions might be taken in warped ways due to the prevailing questionable understanding of stem cells. Abstractions together with limited knowledge and decision-making in the state of scientific uncertainty are unavoidable. However, abstractions and limits of knowledge have to be borne in mind in order that we are 'thinking with abstractions and not obeying to abstractions' (Stengers, 2011).

Acknowledgements

I am most grateful for the support of my co-editors Arne Manzeschke and Christine Hauskeller, whose thoughtful comments on earlier versions of this chapter helped improve it. Any shortcomings are of course my own.

My research was funded by the Deutsche Forschungsgemeinschaft (DFG, German Research Foundation) – Project 254954344/GRK2073. Many thanks go to my colleagues in the DFG research network 2073 *Ethics and Epistemology of Science* at the Universities of Bielefeld and Hannover for the opportunity to discuss earlier versions of the chapter and their helpful comments, especially to Rui de Souza Só Maia and Minea Gartzlaff. This chapter draws on two now completed research projects: the international and interdisciplinary summer school on *Pluripotent Stem Cells*, organized by Arne Manzeschke and me, and funded by the German Ministry of Education and Research (BMBF), and our ethical sub-project in the Bavarian research network on human induced pluripotent stem cells, ForIPS, funded by the Bavarian State Ministry of Science and the Arts (funding code D2-F2412.26).

References

Alvarado, S. A. and Yamanaka, S. (2014). Rethinking differentiation: stem cells, regeneration, and plasticity. *Cell*, 157, pp. 110–115.

Brigandt, I. and Love, A. (2017). Reductionism in Biology. *The Stanford Encyclopedia of Philosophy*, [online]. Available at: https://plato.stanford.edu/archives/spr2017/entries/reduction-biology [Accessed 12 May 2018].

Burns, L. (2009). "You are our only hope": Trading metaphorical "magic bullets" for stem cell "superheroes". *Theoretical Medicine and Bioethics*, 30, pp. 427–442.

Caulfield, T., Sipp, D., Murry, C. E., Daley, C. Q. and Kimmelman, J. (2016). Confronting stem cell hype. *Science*, 352(6287), pp. 776–777.

Dimmeler, S., Ding, S., Rando, T. and Trounson, A. (2014). Translational strategies and challenges in regenerative medicine. *Nature Medicine*, 20(8), 814–821.

Dupré, J. and Nicholson, D. J. (2018). A manifesto for a processual philosophy of biology. In: D. J. Nicholson and J. Dupré, eds., *Everything Flows. Towards a Processual Philosophy of Biology*. Oxford: Oxford University Press, pp. 3–49.

Fagan, M. B. (2013a). *Philosophy of Stem Cell Biology. Knowledge in Flesh and Blood*. London: Palgrave Macmillan.

Fagan, M. B. (2013b). The stem cell uncertainty principle. *Philosophy of Science*, 80, pp. 945–957.

Fagan, M. B. (2017). Stem cell lineages: between cell and organism. *Philosophy, Theory and Practice in Biology*, 9(6), pp. 1–23.

Fehr, C. and Plaisance, K. S. (2010). Socially relevant philosophy of science: an introduction. *Synthese*, 177, pp. 301–316.

Flake, A.W. (2004). The conceptual application of systems theory to stem cell biology: a matter of context. *Blood Cells, Molecules, and Diseases*, 32, pp. 58–64.

Goetz, M. (2018): Revising concepts about adult stem cells. *Science*, 359(6376), pp. 639–640.

Hauskeller, C., ed. (2002). *Humane Stammzellen – therapeutische Optionen, ökonomische Perspektiven, mediale Vermittlung*. Lengerich: Pabst Science Publishers.

Hauskeller, C. and Weber, S. (2011). Framing pluripotency: iPS cells and the framing of stem cell science. *New Genetics and Society*, 30(4), pp. 415–431.

Heinrich, C., Spagnoli, F. M. and Berninger, B. (2015). In vivo reprogramming for tissue repair. *Nature Cell Biology*, 17(3), pp. 204–211.

International Society for Stem Cell Research (2016). Guidelines for Stem Cell Science and Clinical Translation, [online] Available at: www.isscr.org/docs/default-source/all-isscr-guidelines/guidelines-2016/isscr-guidelines-for-stem-cell-research-and-clinical-translation.pdf?sfvrsn=4 [Accessed 12 May 2018].International Society for Stem Cell Research (2008). Patient Handbook on Stem Cell Therapies. [online] Available at: www.closerlookatstemcells.org/docs/default-source/patient-resources/isscr-patient-handbook-english_ltr_17nov2016_web-only.pdf?sfvrsn=2 [Accessed 12 May 2018].

Kaiser, M. I. (2011). The limits of reductionism in the life sciences. *History and Philosophy of the Life Sciences*, 33, pp. 453–476.

Knoepfler, P. (2013). *Stem Cells: An Insider's Guide.* Singapore: World Scientific.

Kraft, A. and Rubin, B. (2016). Changing cells: an analysis of the concept of plasticity in the context of cellular differentiation. *BioSocieties*, 11, pp. 497–525.

Lander, A. D. (2009). The 'stem cell' concept: is it holding us back? *Journal of Biology*, 8, pp. 70.1–70.6.

Laplane, L. (2016). *Cancer Stem Cells: Philosophy and Therapies.* Cambridge: Harvard University Press.

Leychkis, Y., Munzer, S. and Richardson, J. L. (2009). What Is Stemness? *Studies in History and Philosophy of Biological and Biomedical Sciences* Part C, 40(4), pp. 312–320.

Lysaght, T. and Campbell, A. (2013). Broadening the scope of debates around stem cell research. *Bioethics*, 27(5), pp. 251–256.

Maienschein, J., Sunderland, M., Ankeny, R. A. and Robert, J. S. (2008). The Ethos and Ethics of Translational Research. *The American Journal of Bioethics*, 8(3), pp. 43–51.

Nature Editorial (7 July 2016). *False Assumptions. US Regulators Must Regain the Upper Hand in the Approval System for Stem-Cell Treatments*, [online] Vol. 535. Available at: www.nature.com/news/fda-should-stand-firm-on-stem-cell-treatments-1.20208 [Accessed 12 May 2018].

Rheinberger, H.-J. (2006). *Epistemologie des Konkreten. Studien zur Geschichte der modernen Biologie.* Frankfurt am Main: Suhrkamp.

Robert, J. S. (2004). Model systems in stem cell biology. *BioEssays*, 26, pp. 1005–1012.

Robert, J. S., Maienschein, J. and Laubichler, M. (2006). Systems bioethics and stem cell biology. *Journal of Bioethical Inquiry*, 3, pp. 19–31.

Stegenga, J. (2018). *Medical Nihilism.* Oxford: Oxford University Press.

Stengers, I. (2011). "Another science is possible!" A plea for slow science. Inaugural lecture Chair Willy Calewaert 2011–2012 (VUB), [online] Available at: https://de.scribd.com/document/109374172/Another-science-is-possible-A-plea-for-slow-science-Isabelle-Stengers-2011 [Accessed 12 May 2018].

Tajbakhsh, S. (2009). What's in a Name? *Nature Reports Stem Cell*, [online] Available at: www.nature.com/stemcells/2009/0906/090625/full/stemcells.2009.90.html [Accessed 12 May 2018].

Towns, C. R. and Jones, D. G. (2004). Stem cells, embryos and the environment: a context for both science and ethics. *Journal of Medical Ethics*, 30, pp. 410–413.

Trounson, A. and De Witt, N. D. (2016). Pluripotent Stem Cells Progressing to the Clinic. *Nature Review Molecular Cell Biology*, 17, pp. 194–200.

Turner, L. and Knoepfler, P. (2016). Selling Stem Cells in the USA: Assessing the Direct-to-Consumer Industry. *Cell Stem Cell*, 19(2), pp. 154–157.

Zipori, D. (2004). Opinion: The nature of stem cells: state rather than entity. *Nature Reviews Genetics*, 5, pp. 873–878.

10 Mapping laboratories and pinpointing intentions

The entanglement of audit and reproducibility in the STAP case

Melpomeni Antonakaki

Abstract

This chapter revisits a recent scientific controversy, which debated matters of quality-control and credibility of published research on an experimental hypothesis. The case at hand often goes by the name of the proposed hypothesis itself, namely STAP, the acronym for Stimulus Triggered Acquisition of Pluripotency, a hypothesis situated in the field of stem cell biotechnologies. By closely studying the official reports, follow-up publications, and auditing processes related to the accused authors and the employer organization, I attempt here a critical analysis of the investigation processes that argued back in late 2014 to have resolved the case as one of misconduct on behalf of the principal author. My research has revealed a much more complex outcome of said investigation, which only when severely amputated and selectively reproduced can be called a fair or true resolution. Staying with the troubles and labours for building a misconduct case at a major research organization in Japan, this chapter brings forth the entanglement of audit and reproducibility not only as observed in this controversy, but crucially as framed and advocated for in the global desiderata and guidelines concerning misconduct in (bio)science.

Introduction: situating 'fraud' within controversy

This chapter revisits a recent scientific controversy in the field of stem cell biotechnologies, which broke out in Japan in early 2014 following the publication of two scholarly reports (under the categories of 'research article' and 'letter') at a highly respectable journal. The 'fruits' of an esteemed collaboration between Japanese and American experts in the field, the reports argued for a breakthrough method for generating cells which exhibit markers of pluripotency (Obokata et al., 2014a, 2014b; both papers have been retracted). The method had been called Stimulus Triggered Acquisition of Pluripotency (STAP). But no later than a week upon the online appearance of the two reports there arose allegations over the quality of the published findings, even doubts over the reproducibility of the STAP method itself. The allegations were mainly directed against the female 'corresponding author', but they also targeted her famous and established supervisors. An official inquiry over the credibility of the allegations was instigated almost immediately from the Japanese side, and for almost a year all involved parties underwent waves of intensified scrutiny. The *STAP case*, as it has come to be known since, has been widely reported in the media as yet another stem cell-related fraud, often compared to the Hwang case in South Korea (Goodyear, 2016; Martin, 2014; Normile and Vogel, 2014; Van Noorden, 2014; Vogel and Normile, 2014). Especially those media platforms that relate to scientific issues or communities have attracted since then the attention of social scientists who in turn diagnose the emergence of a novel culture for debating *matters of conduct* in scientific research along the public – sometimes stereotyped – image of institutions and even whole nations (Gayle and Shimaoka, 2017; Lancaster, 2016; Meskus et al., 2017).

In this chapter I offer an analysis which highlights features of institutional processes of auditing and the related in-house design of reproducibility attempts (the not-so-obvious elements of the STAP case), and find both to operate as targeted and effective methods for preserving institutional survival and status protection at the cost of individual scientists' reputation and careers. My approach is inspired by the field of Science and Technology Studies (STS) and its commitment to treat *controversies* as cases *par excellence* for analytically attending to the social and political life of scientific methods. My chapter contributes to this line of research with a theoretical argument which suggests the entanglement of auditing routines and reproducibility settings in contemporary stem cell biotechnologies. I speak of 'entanglement' drawing on Karen Barad's work, to argue not simply for an empirically observed 'intertwining of two (or more) states/entities/events, but a calling into question of the very nature of two-ness', their ordering separate jurisdictions – one of institutional control over employees, one of peer-assessment of skill and technique (Barad, 2010, 251). Studying the organizational and – beyond institutional borders – social life of the misconduct investigation in the STAP case, I show how auditing methods articulate multiple *experimental* questions in the biological laboratory as much as how biological experiments may dictate the means and objectives for the proliferation of institutional audit objectives and mechanisms. In shifting the perspective on the case from the scandalous to its more conventional and epistemic aspects, I treat the STAP episode less as an 'anecdote' and more as a paradigmatic case – one found situated at the heart of epistemic and institutional debates that the stem cell biotech field currently faces regarding reproducibility and data/records quality. Thus the case can be approached as highlighting a more general need to come up with 'a new sense of a-count-ability, a new arithmetic, a new calculus of response/ability' in order to attend to pressing issues regarding (mis)conduct in scientific practice and authorship (ibid.).

My argument has two parts. First I explore the literal and literary grounds of investigation, starting with their constitution and meaningful mobilization within the investigatory agenda. Each subchapter presents different phases of the investigation process and evaluates their individual performance and ethos in a reading against the official accounts and reports. Part II is concerned with the broader logic and working framework that provides supra-organizational grounds for legitimacy of such types of investigation, namely the Global Science Forum's (GSF's) 'quasi-legal' approach to misconduct investigation. This move, which necessarily escapes the Japanese institute's borders and jurisdiction, is crucial for legitimizing certain designs and choices within the logic put forth by global desiderata and guidelines.

A note on the material I draw from: the regular disclosure of official documentation on the STAP protagonists resulted in the case constantly featuring among the popular bioscience news of 2014 and thus multiple versions of the case were generated and circulated in physical and virtual spaces of popular media. Although the role of media attention has been extremely dominant in the public communication of the case, I do not draw here on media-generated or supported discourse as I seek to uncover the concepts and objectives guiding the actual in-house investigation. As will become obvious, the official reports are highly technical documents. Neither their dry, reporting tone of language, nor the uneventful details and conclusions would have met the requirements of the sensationalized piece of news the media sought. To a great extent the investigation outcomes were superficially reproduced in the media and have remained black-boxed until now. To counterbalance the invisibilities and injustices the STAP case protagonists endured against the aggressive spectacle that was manufactured on the misfortunes of many, I dedicated this chapter to the 'boring stuff' (cf. Star, 1999).

Part I: manifold accounts of the STAP investigation

The responsibility for examining the allegations against the two Obokata et al. publications was taken on by the employer organization of four out of the five main co-authors (the famous Center for Developmental Biology, or CDB of the Japanese institute RIKEN, which has however been renamed). We learn from the introductory statement of the first published report that the action took place immediately after one 'RIKEN researcher who had been notified of doubts concerning research papers published by RIKEN scientists contacted the RIKEN Auditing and Compliance Office through one of RIKEN's executive officers' (see 'circumstances', in Ishii et al., 2014, 1). In this sentence the repeated emphasis on the name of the employer entity is noteworthy. At that point it highlighted the organizational grounds on which all elements in the chain of action (from notification to reception) took place and it precipitated the chain of command to follow. At a single stroke the sentence was enacting RIKEN as the sole responsible actor to take action. Since its inception the STAP investigation was constituted as a matter of RIKEN's administration, *an administrative response* to affairs made 'internal'.

This *internalization* is paramount for a reason. RIKEN struggled through-out 2014 to contain the narrative on the STAP case by continuously assuming responsibility for intended or unintended (re)actions to the outcomes of the investigation. I argue that such struggle for the containment of accounts is crucial for identifying/analysing institutional accountability structures and assess the role of gestures of transparency. Both said structures and gestures allow for identifying unique grids of knowledge/power, and offer a rare opportunity for the study of unequal scales of authority and credibility. For the best part of eleven months (February–December 2014), the investigation over the STAP case spread across diverse expert 'panels' (some internal, some external, mostly comprised of Japanese biologists) but always on CDB's grounds and within the RIKEN regulatory framework. A whole range of technical administrative procedures, experiments, and interrogation strategies were mobilized for the sake of generating knowledge-based evidence of misconduct.

It is also important to differentiate between processes that held Obokata et al. accountable to RIKEN, and RIKEN accountable to the Japanese state and broader public(s). My analysis focusses on how the boundaries were drawn and maintained between internal RIKEN affairs and the out-side world, in order to stage the auditing of certain employees and practices as an *outward looking* yet organizationally contained enterprise. To address the branching out movement of auditing methods while also tracing how the investigation reassembles itself as a unified enterprise, I situate the reader inside the *topos of investigation*, a geography of RIKEN facilities emerging from the documents and images I have studied. Such staging of the STAP episode follows time frames and spatial arrangements relevant to people or panels involved in the STAP case, 'acting' either in their capacity as authors of audit reports or as subjects/suspects of investigation. I have drawn from and analysed the whole volume of investigation-related documents, which were publicly disclosed for the sake of transparency over the selected action(s) and decisions. My methods also included the analysis of certain scientific publications and policy documents that were directly related to or built further upon the investigation-practices and experiments around the STAP case.

Investigating scientific authorship

A panel of distinguished, mostly RIKEN-based scientists was formed and began their inquiries into the details of the case on 13 February 2014. As soon as their preliminary research (13–17 February 2014) suggested that there were grounds for the launch of a full investigation, the same panel, featuring an extra member (a lawyer), was renamed to 'investigative committee' (hereafter, 'committee') and started collecting all material related to the STAP publications for critical examination. They specifically asked for all original data that could be assembled from laboratory notes, early drafts, emails exchanged among the four employees under investigation, and other 'supporting' written material, such as information on the equipment used in the experiments. The committee focussed on collecting as many layers of writing and copy-editing samples as possible. The examination of this material included consulting an imaging expert and structuring interviews with the four people of interest: lead and corresponding author H. Obokata, second corresponding author Y. Sasai, third corresponding author T. Wakayama, and co-author H. Niwa (ibid., 2).

Laboratory notebooks and early drafts were collected as evidence and compared to one another, accounting in this manner for their authors' commitment to RIKEN's auditing requirements. This focus on written text/ notes renders auditing practices in technoscience a hybrid genre of authorship within which unequal scales and qualities of knowledge and authority clash. Humanities scholars have noted how laboratory notebooks and digital platforms for record-keeping have been granted a special status within official audit requirements in public-funded research (Neyland and Woolgar, 2002; Shore, 2008; Strathern, 1997). Compulsory record-keeping often figures as a practice that structures everyday routines in 'the lab' for the sake of researchers and institutions, and – at the same time – as providing auditors with invaluable raw insight when extraordinary situations arise (e.g. misconduct allegations). Daniel Neyland has specifically addressed such manifold performance of audit practices through the concept of *parasitism*, according to which counting, accounting for, and holding accountable often coincide (within texts and organizational spaces), thereby proliferating written accounts and generating more audit for the sake of audit (2012). The initial phase of the STAP investigation provides a rare window into how these manifold requirements operate on those subjected to audit: the parasitic (or productive) capacity of auditing makes *authors* out of those involved in investigation, both auditors and suspects. The people implicated in the STAP publications were questioned for the texts they had authored, whilst simultaneously the investigators themselves became authors when translating a body of samples into evidence of character or fault.

'Evidence' was sought either in the continuity or in the discrepancies between samples of authorship. In its initial report dated 31 March 2014 (Ishii et al., 2014), the investigation identified two misconduct instances and several minor problems in the writing procedure of the papers. Specifically, at the process of transferring data from laboratory notes into the text of scientific articles. Obokata was found responsible for 'manipulating the image data of two different gels and using data from two different experiments' (ibid., 9). The manipulation of an image fell under the definition of falsification as found on page five of the same report, and mixed data from different experiments fell under the definition of fabrication on page eight. Yet the status of other allegations remained opaque, because they did not fit the definitions of falsification, fabrication, or plagiarism.

Interestingly, the report continued with a discussion specifically of the person of Obokata. The committee expressed an overall assessment of her character as sloppy, 'lack[ing], not only a sense of research ethics, but also integrity and humility as a scientific researcher' (ibid.). Paramount to this conclusion is the state of her examined laboratory records: '[g]iven the poor quality of her laboratory notes it has become evident that it will be extremely difficult for anyone else to accurately trace or understand her experiments' (ibid.). Notably similar assessments of the personality of other authors are not included in the report. Instead the authors held the other interviewees responsible as failed supervisors, thereby diagnosing a major 'failure of the normal [audit] system of checking on research results' (ibid.).

On the morning of 1 April 2014, RIKEN held a press conference on the STAP case, which focussed on how the company planned to regain public trust on two separate fronts: on the one hand, Obokata's person was publicly discredited as a sloppy researcher, a move that strategically separated her person from the STAP method; and on the other hand, the failure of the 'normal system' (as detected by the committee) was promised to be addressed in a separate auditing, this time relating to the CDB as a whole. Overall, the statement dictated the means by which investigation would continue (verification of STAP method, appeal process for Obokata, disciplinary measures for the parties involved) by reinforcing an adamant belief in scientific ethos and by streamlining an 'independent audit' concerning the reform of the internal governance mechanism (MEXT, 2015, 11–14). The main question at RIKEN's 'public accountability' performance in April 2014 can be interrogatively summarized as follows: Obokata aside, what structural elements made the STAP case a liability issue? This move is a re-localization of responsibility concerns from the individualized to the systemic and it is very interesting indeed. To echo Power's argument concerning the explosion of audit, and Pels' affirmation of this thesis, the main role of audit is to control the control, to police the policing; and in RIKEN's case, to review the prescribed system against its own failure (Pels, 2000, 141; Power, 1994, 6).

Up to a certain degree, the decision to scale up the reform of auditing regulations backfired for RIKEN after April 2014. By 12 June, the newly assembled 'Reform Promotion Headquarters for preventing recurrence of research fraudulent' and an external reform committee had reached the conclusion that there was, among other issues, a total 'absence of a practical system to record and manage experimental data' (MEXT, 2014, 11). In the light of such findings, what was perceived as a singular episode of misconduct (the system failure in the Obokata et al. case) became writing-as-usual within RIKEN, and as far as the data- and record-keeping economy was concerned, the line separating proper- from sloppy-writing skill became more difficult to draw. Looking back one might ask whether it is the failing system to blame for the STAP misfortunes or the individual users of said system. Such questions did indeed arise within RIKEN in 2014 but were confronted as parts of an enterprise separate from the STAP case – namely the reorganization of the CDB and the reformed regulatory framework. One of the outcomes of the detected separation was that Obokata was afforded a very slight chance to actually recover from the pejorative characterizations (ascribed in the first committee's report) which went viral across the media.

The circle of discredit in the Appeal process

It is worth looking more closely at the methods that immediately relate to audit and its focus on authorship, not only for the devastating effects it had on Obokata's reputation but in order to problematize the structure and role of the interviews conducted between February and April 2014. Reading an audit through its own narrative devices permits one to escape a certain analytic trap: although the etymology of the word 'audit' would suggest its mobilization of a 'close hearing' of testimony, Marilyn Strathern has argued instead that one ought to pay attention to its constant inventing of accountability which operates as a powerful technology of vision and focus (Strathern, 1997). The over-exposure of testimony deemed proper and relevant induces a simultaneous invisibility of alternative accounts. For social studies of technoscience this constitutes a constant challenge so to avoid the blind spots of audit authorities or audit authorship, and to seek out the accounts rendered irrelevant or improper at the periphery of this directed gaze (Garforth, 2012). For the STAP case it is of importance to analyse the audit documentation along two dimensions: as an attempt to gain an understanding of the nature of formal inquiry and as a means of providing insight into the content and character of interview sessions.

Only days after the disclosure of the committee's report, Obokata strongly opposed its conclusions in a formal appeal process (hereafter, *Appeal*). The Appeal targeted the investigatory methods deployed by the committee, the structuring of the interviews, and the brief space of time she was given to prepare her defence. Interestingly, one learns the above features of the Appeal through the documents that report its rejection. In the Report on Review of Appeal (RoA) one witnesses the formation of a circle of discredit: the committee was asked to review its previous findings in the light of Obokata's criticisms, and they proceeded by discrediting her arguments in the light of their findings in the report (Watanabe et al., 2014).

An example of how this circular discursive strategy worked emerges by focussing on how Obokata's statements during the first period are incorporated in the context of the RoA. The materials that appear as Obokata's statements include speech acts of different qualities (some are quotations taken from the Appeal's text-body, other are quotations of free-flowing speech from interview transcripts, some are excerpts from the investigators' notes on what she said, some are a mixture of all these).[1] Consequently, what here I call 'Obokata' is the assemblage of different references, a copy-edited persona which only emerges within this discursive circle of discredit. The extensive parts that claim to be from the Appeal and the Appeal's supplementary documentation (both were never made public) were presented in a way that contradicts the Obokata-of-the-notes taken by the interviewers (e.g. on demonstrating how she was knowledgeably misleading them, see Watanabe et al., 2014, 4–5). By the skilful juxtaposition of 'her' speech acts, an inconsistent persona is generated.

The committee's report from 31 March 2014 mentions that several interviews were conducted and it summarized the findings, but nothing was included on the chosen structure of these interviews, on the questions posed, nor on their number/frequency and setting. An insight into the organizational details that were not included in the committee's reports, but which are of relevance to this paper, can be gained by observing the many layers of material pertaining to the STAP case that the RoA offers. This material is invaluable, no matter its detected argumentative circle of discrediting one interviewee. The inconsistent 'Obokata' comes into existence only if and when the same question is posed repeatedly throughout a number of interviews and thereafter assembling and reading the responses one against the other (for the most striking schema of discrediting, see the paragraph, 'On February ... visual inspection', in ibid., 5). The text was actually reviewing the Appeal by reading it against Obokata's person, and not against the panel's workings and methods, which constituted the actual target of the Appeal process. This was an incredibly effective strategy: at the end, she (the literary persona) confirms their initial findings of being sloppy, and she (the person appealing) is rejected on the grounds of her own faults.

This short period of Reviewing the Appeal during April 2014 enabled a change of focus for the investigation as a whole. Hitherto, the authors of STAP were coextensive with the textual testimony provided by the written accounts. From that point onwards inconsistencies of statements became a relevant object of investigation and questions of misconduct exited the space of investigating authorship and entered the laboratory. When the juxtaposition of multiple answers showed the appellant to have 'some awareness of the result', the issue of misconduct became a matter of personal awareness, a conclusion that went much further than sloppiness (ibid.), reaching into the space of laboratory practice.

Establishing strong intentionality

The sum of decisions, people, data, methods, and texts that carried the STAP investigation up to November 2014 was set into motion by an objective to pinpoint fraudulent intentions to the persons under investigation using scientific means. This objective materialized in three intertwined experimental settings within RIKEN: the verification laboratory, the scientific validity examination team, and the research publication investigative panel. Verification is taken here to signify a demonstration of evidence in support of the STAP method and for the sake of the organizational reputation. It is not a peer-reviewed practice or reproducibility effort, and it comes with institutional and personal stakes that were made very explicit starting with

April's public statement. The examination setting is also interesting as it was designed as a process parallel to the previous one: a comparative experiment with its own objectives. To grasp its peculiarity one has to abandon a conventional understanding of reproducibility. Finally, the last panel emerges from my analysis as the fail-safe mechanism for resolving the case, put there as Plan B in case the previous attempts bore no fruit. To reconstruct in a few paragraphs the full spectrum of experiment-based investigation is a challenging task. My aim is to present briefly all three settings and discuss the overall emerging mapping of laboratories and intentionalities.

The actual conditions of verification experiments and Obokata's presence at RIKEN were for many months the subject of media speculation. Only after an official scientific report – in the form of a prepublication draft – was uploaded at the online platform bioRxiv, were certain details about the endeavour of verification made public (Aizawa, 2015). It stated that Obokata 'was permitted to perform this closely monitored study on the 14th of July' (ibid., 2). The closely monitored character of the study included getting her to reproduce the very first step of the STAP method (namely, giving the acid bath, and observing the cell phase with fluorescent microscopy) in a newly furnished laboratory, with brand new equipment, without permitting her access to her old bench and facilities. She was only allowed to move in designated rooms, and she had to work under surveillance from both cameras and expert witnesses (ibid., section 'Acknowledgements'). Dr Aizawa, as the person in charge and author of the report, encouraged readers to recognize a series of limitations in these experiments. A specific co-author's presence for example would have been paramount for sticking to the specificity of the STAP protocol – especially regarding chimera formation – but he had refused participation. Moreover, the experiments 'were conducted under strict time constraints, and in the face of considerable, often adversarial media scrutiny' (ibid., 7).

Simultaneously, the validity examination team, led by co-author Niwa, was repeating the STAP protocol, while also receiving and checking her cell aggregates for analysis (Niwa, 2015). This set-up, based on RIKEN premises and staged in close physical proximity to Obokata's new laboratory, is understood here as pertaining to a matrix of clashing rationales: it is at once part of *Verification, Investigation,* and *Reproducibility.* Its parallel structure is found to complement the verification of method, by continuing the steps of the protocol beyond aggregate formation; secondly, it investigates the quality of Obokata's outcomes of practice; and, last but not least, it substitutes for the aforementioned absence of one co-author. In order to take part in such efforts, two of the authors of STAP publications were admitted back into laboratories and judged as practitioners of science. Another aspect of the parallelism of laboratory work by Obokata and Niwa in these efforts testifies to an asymmetry of placement. Undoubtedly, they were both challenged to perform successfully, with institutional and individual reputations at stake. Still, Obokata's position is far worse. What else can one infer from the surveillance imposed, the limitations regarding her participation and movements, and the decision that the results would not be disclosed to her? The set-up and distribution of work illustrates that mistrust against her person was constitutional for the verification process.

On 3 September 2014 an independent panel of external experts was established as an immediate response by RIKEN to the announcement of zero trace of STAP cell-like clusters in the interim report of 27 August. The Katsura et al. panel conferred for a span of eight weeks. They were searching for viable scenarios of compromised cell identity (a.k.a. contamination), which could become the basis of further investigating the misconduct allegations. The scenarios of contamination (imagined as cell circulation and mixed communities of cells) were based on an experiment in comparative genomics, an information-based approach for ordering and sorting cell line variation in stem cell technologies. In this case this meant checking and comparing the genomic information of other cell lines against the characteristics of the STAP lines to find connections and similarities.

The Katsura team reported extensively (but not exhaustively) on the steps taken in all three scenarios of genomic matching. Their semi-scientific, semi-investigatory report is a very peculiar text which allows for the reconstruction of this strategy (Katsura et al., 2014). Establishing a strong genomic link between two cell populations requires demonstration of extended similarity. In the case in point, what could work as an experimental chain-of-proof would be the detection of the same inserted DNA sequence, which when combined with other characteristic deletions in the genome may then indicate a cellular or genealogical relation. But genomic make-up similarity is not enough. The initial hit must be complemented by further data on the type and/or sex attributed to the cell lines, and by the date of cell line establishment and technical nomenclature in order to introduce facts relevant to the laboratory geography and its coming-and-going personnel into the question. For example, to support the hypothesis of extensive contamination, the result had to demonstrate that the amount of genomic variation was not suggestive of cells isolated from lab-mice bodies, meaning that the contamination did not take place during chimera formation. Showing that the STAP cells were descendants of other cell lines, the investigators were able then at another step of interrogation to ask explicitly whether those cells had been transferred or added manually in the mix-up at initial steps in the STAP experiment.

In the end three contamination hypotheses were drawn out, each pertaining to an arithmetic of laboratory events and trajectories of purposeful movement: cell circulation among three different laboratories hinted at either accidental or intentional insertion of representatives of an embryonic cell line to the Petri dishes, vials and test tubes, Eppendorf tubes, fridges, data sets, etc., of the STAPs. The cell lines identified as STAP-related, as orderly organized in a table on page three of the Katsura et al. Report, constitute an imaginative proliferation of relations and connections in the STAP case. Connections between human, technical (machine and utensil), cellular

(STAP, fertilized egg embryonic stem cells, unknowns, nuclear transfer stem cells, samples from teratomas), and animal (parental mice strains and chimera mice) actors. Moreover, these connections are reconfigured along the space and time frames that are meaningful to auditing: namely data pooled from tags, data sets, laboratory records, and archives that belonged to three specific facilities. This confirms how human movements and intentions were the subject of the 'genomics' experiment all along, because the genomic tests focussed only on places to which the authors Obokata and Wakayama had access, where they stored material, etc. Interestingly this made temporary 'suspects' of all laboratory personnel using those rooms.

This kind of strategy does not reflect conventional reproducibility methods. The Katsura report shows that the principle guiding their work does not map along the usual categories of recapitulating settings and replicating results. Findings on cellular relation do not merit much speculation while they remain at the level of information/data set. What matters is the effort made in translating the knowledge generated by genomic findings into a template for imagining the practice of misconduct. I argue therefore that it was the possibility of mapping intentional movements of cells (and pointing fingers at the humans responsible for them) which designed experiments on cell identity in the first place. In the aftermath of producing these new insights on the STAP case, the investigators structured a series of new interviews with Obokata and Niwa, where the genomic findings were further used as interrogation material. While the contamination-inspired interrogation was unfolding, over a span of eight weeks and fifteen meetings until mid-December 2014, two out of the three persons of interest (Obokata and Niwa) were involved daily in the verification and validity experiments; and while in the process of verifying the STAP hypothesis, they were also subject to interrogation that aimed at proving that the STAP method was either fraud or accident all along.

The last investigatory panel built an unusual setting for investigating accountability and compliance issues within a scientific institute. They pulled together a hybrid of science and auditing mechanisms: generalizing on who counts as a suspect, recruiting personnel from other facilities, and re-enacting the 'crime scene' via biological means. One needs to acknowledge the hard work and involvement of several scientists and RIKEN employees (at least two Attorneys of Law, personnel from seven laboratories, and a large number of other people, several of whom were undergoing restructuring of their facilities within CDB or were at the cusp of retiring) (Katsura et al., 2014, 33). At the same time, acknowledging the innovation of approach does not equal overlooking the outcome. This ambitious and imaginative setting failed to deliver its promise of establishing intent of misconduct, especially since it never delivered any confession of guilt. All audit and reproducibility efforts came to a sudden halt around 20 November 2014, after the third panel supported the contamination-in-great-extent hypothesis, even though verification attempts were designed to reach March 2015 (Normile and Vogel, 2014).

There is an inconsistency in the public documents as to what exactly Obokata's status in RIKEN was by late November 2014, with MEXT's review stating that she left the verification experiments on 30 November (MEXT, 2015, 13) and the investigatory panel reporting that she lost her Unit Leader status on November 20 and was transferred to the Office for the Prevention of Research Misconduct until December 21 (Katsura et al., 2014, 2).

Wrapping up the topos of investigation

RIKEN's 2014 full-blown investigation was a complex enterprise and involved many different actors from within and outside the organization. It may have begun as the implementation of the standard response mechanism of the Office of Auditing and Compliance, but during its course it catalysed major transformations of RIKEN's audit and governance system, and many outcomes of this scaling-up were unforeseen. In the STAP case, panels of experts were tested for their ability to prove misconduct and fraud in scientific terms. Each adopted a different focus and deployed methodology deemed suitable; yet taken as a whole the succession of findings does not follow a linear narrative and the findings are inconclusive to a great extent. Moreover, the branching out movement of investigation towards reproducibility questions featured incompatible agendas in the form of incompatible experimental questions: the verification was premised on Obokata performing successfully the acid test; the validity team, that others than Obokata could perform successfully the whole sequence of STAP protocol. And the Katsura team was based on the premise that no one would, because the STAP never existed as such.

The full body of investigation methods was geared towards accelerating a sequence of findings that discredited specifically Obokata's person: she was deemed sloppy and untrustworthy at the end of the first round of investigation, inconsistent and pending disciplinary measures after the rejection of the Appeal, an institutional asset when she returned to the laboratory practice, and finally a failure when she failed to create new STAP aggregates. Considering the manifold performances and the vigour of the investigation as a whole, it is remarkable that by the end of 2014 all aforementioned projects had managed to add only two more instances of Obokata's misconduct (in relation again to documentation) to the first two. In the meantime, the effects of the investigation and the publicity that accompanied it had been devastating for many: not only did Obokata find herself committed twice, but she was also physically injured and housebound for months; the head of the first committee had to resign under the weight of STAP-unrelated, yet misconduct-related, allegations; Obokata's supervisor and fourth co-author committed suicide on RIKEN grounds; the whole CDB underwent restructuring. Moreover, both professors Wakayama and Niwa were demoted in institutional status; and the chain of steps and events expanded beyond RIKEN, with Japanese universities investigating doctoral dissertations and reforming their audit regulations (Kobayashi, 2016).

The sum of these effects confronts one with aporias related to the creation of legitimacy and the entangled performance of these specific investigatory choices, which I discuss as such in the next section. The material analysed in the first part of the chapter becomes the basis of inquiry for the history and situatedness of misconduct investigation policies.

Part II: the means and investigative tropes of quasi-legality

[The misconduct investigation process] has many of the attributes of criminal or civil procedures, but is reduced in complexity and is meant to function more quickly.

(Global Science Forum, 2007)

The many shortcomings of the STAP investigation as analysed so far raise a series of questions concerning the logic which rendered the actions and decisions taken in the STAP case as a coherent and legitimate agenda – especially if one considers how small scale and far from media attention other audit investigations had been in the same country (even in the same organization) (cf. MEXT, 2015). The question is what makes certain strategies legitimate responses to misconduct allegations and what 'factors' dictate how and when the issue becomes resolved. The main international authoritative source regarding this problem is the GSF's agenda of quasi-legal scientific investigations of misconduct. The series of texts on *Research Integrity* that gave rise to this framework seem to be the basis for MEXT and RIKEN regulations (the period 2007–2014), as becomes obvious when one follows the cross-referential economy of the texts. Furthermore, the actual steps of investigation correspond closely to the suggestions found therein (analysis below). Interestingly both MEXT and RIKEN have revised their guidelines in the aftermath of the STAP case in a way that makes it more difficult to debate the relation between global desiderata and the current framework.

The STAP case offers a unique window into a more detailed understanding of such policies. Revisiting it, I am interested in how the institutional frameworks in this specific country and this specific institute have been situating themselves among global actors and desiderata. To this end, I first focus on how investigatory processes regarding misconduct have been agreed upon within the politicized context of the Lisbon Conference and the Tokyo Report, in order to develop an understanding of the drawbacks and dead ends of experimenting and auditing inspired by the so-called *quasi-legal* approach to misconduct investigations.

Sticking to the book: a to-do list for investigators/scientists

The first World Conference on Research Integrity took place in Lisbon, Portugal, in mid-September 2007 with discussion on the so-called Integrity approach. Delegations from two countries – Japan and Canada – jointly initiated this process, by organizing the Tokyo workshop of 'Best Practices for Ensuring Scientific Integrity and Preventing Misconduct' in late February 2007 (Global Science Forum, 2007). The Tokyo meeting was convened to provide the GSF with a first outline of a report that could form the basis for the World Conference. The Integrity approach draft proposal (hereafter, the *Tokyo report*) is based on a text composed after discussing strategies and shared definitions for harmonizing misconduct policies, in order to forge consensus on the character of legitimate intervention (Global Science Forum, 2007). The Research Integrity Approach was built on the idea that the scientific community in the very strict sense constitutes the expertise *par excellence* to deal with what is considered an aberration from the scientific canon of conduct. Very early in the Tokyo report we find that misconduct is an attack directed to science itself, 'by creating false leads for other scientists to follow, and/ or forcing others to waste time, effort and money to reproduce fraudulent results' (ibid., 4). Wasting money and time are crucial elements of replication/ verification attempts that go sour.[2] But as the report continues, '[f]ortunately, the research record is inherently self-correcting, since repeatability, verifiability and consistency are hallmarks of the scientific method' (ibid.).

The document also tried to redress and transgress cultural and/or national discourses of research-related credible practices and recognizes cultural and institutional diversity and/or freedom to act accordingly. The report's aim was to incorporate case-specific particularities into a quasi-legal convention that works in all contexts by establishing intent (ibid., 4) and fairness (ibid., 6–9). In this framework, quasi-legality (an approximation of a system that serves and delivers justice) constitutes the universal foundation for national agendas, while intentionality and individual rights and obligations are the building blocks of the process. The tools for national initiatives to raise their super-structure of guidelines are 'prevention' and 'enforcement' (ibid., 2, 12). The convention seeks to develop common workable definitions, suggesting that an *administrative understanding of investigation* be adopted by esteemed representatives of science and, explicitly, without burdening external regulatory bodies (ibid., 9). The world conference addressed a much broader spectrum of stakeholders in science, bringing together 'researchers, research administrators, research sponsors, journal editors, representatives from professional societies, policy-makers, and others', although its conclusions excluded regulatory science and related bodies (Mayer and Steneck, 2007, 1).

Overall the logic of Research Integrity explicitly aimed at being descriptive, and not normative. It is an orienting framework and so its adoption or implementation should be understood as flexible. The manual-like structure of the Tokyo report highlights the conviction of OECD for *descriptive suggestions*. It concerns the deployment of a regulatory framework in the form not of a recipe for action (with steps to be followed), but a *to-do list* – questions to be asked, areas to be covered, things to be kept in mind – in order to achieve investigation(s) across cultures and organizations. It claims to be of an informative and advisory nature (Global Science Forum, 2007, 2), and proceeds with the following sections: Varieties of Misconduct (p. 2), Options for Dealing with Research Misconduct Allegations (p. 5), Responding to Misconduct (p. 8), Investigating Misconduct (p. 9), International Considerations (p. 11), and Prevention (p. 11). The paragraphs concerning Responding and Investigating Misconduct consist of lists of questions per topic.

The above-described texts, which outline the quasi-legality of misconduct investigations, contribute to understanding RIKEN's decision-making in the wake of the STAP-related allegations, and throughout 2014. The almost one-to-one detected correspondence between policy suggestions and the panel's actual design of the investigation is worth pointing out. But one needs to imagine this in the form of a Q&A session: one side sets the questions, one gives answers to them. A detailed presentation of how the to-do suggestions and case-study specific actions correspond would have exceeded the scope of this paper and a sole example has to suffice (Global Science Forum, 2007, 9). The next section on the legitimacy of quasi legality can be read as RIKEN's or investigators' *answers* to the fifth bullet point of investigating misconduct: 'Are there limits on the power/authority of the investigators? How much new work can they require (for example, repeating an entire series of experiments)?' (Ibid., 9) The partnering of the question on the limitations of authority with the limitations of available resources reflects the orienting yet open-ended character of the quasi-legal to-do list.

Quasi-legal configurations of knowledge, truth, and justice

To explore the grounds for the legitimacy of the quasi-legal approach, I begin with its frame as presented in the text:

> Questions of fairness are particularly important when dealing with misconduct, because the investigation process has many of the attributes of criminal or civil procedures .... Precise definitions, policies and procedures for misconduct investigations are needed to prevent the perception (or, worse, the reality) of a "witch hunt".
>
> (Global Science Forum, 2007, 9)

Quasi-legality is enacted by bringing a sort of legal expertise into the courts of science, in an effort to produce and control non-legal, yet intentional subjects. In the Tokyo report, intentionality emerges as a personal trait, yet what is at stake in an 'optimal mapping' and 'optimal response' is its establishment as an epistemic object. At its core this logic is the opposite of what Foucault described as modernity's punitive reason: instead of the Law recruiting scientific expertise to produce the criminal mind and body, we see Science re-enacted as a sort of court or trial, inserting elements of legal expertise to produce fraudulent intentions as an epistemic event (Foucault, 1995). A genealogy of practices pertaining to this productive reversal has been the focus of a series of lectures by the late Foucault ([1981] 2014).

There the philosopher analyses the tropes and means of 'veridiction' (speech acts pertaining to the jurisdiction of truth) as 'weapon[s] in relationships between individuals, ... as means of modifying relations of power among those who speak, and finally as element[s] within an institutional structure' (ibid., 28). It is on this jurisdiction of truth-finding and truth-telling that quasi-legality claims its territory and here, too, the scientists emerge as the sole expert community for misconduct investigation.

In the Tokyo report the term *quasi-legal* is taken to signify both fair prosecution and precise language. The specific deployment of such convictions in the STAP case is telling of the limitations and inherent violence of the quasi-legal format. The constitutive move for legitimizing the grounds and early results of investigation was indeed a legal term: *akui* (悪意). *Akui* was spotted by the first investigatory committee in RIKEN's audit regulations, and inserted in their discrediting strategy at the Report on Review of Obokata's Appeal. Note that the specific notion means 'with malicious intent' in everyday use, but 'knowingly' in the legal sense (Mizuno and Nakamura, 2009, 13 [194]). As courtroom discourse analyst Nakamura and court interpreter Mizuno point out 'legal terminology in Japan is highly specialized and difficult for lay people to understand', as is the case for most countries, and *Akui* is an example of legal jargon triggering courtroom incommensurabilities between lay and legal communication (in translating, interpreting, interrogating) due to the negative charge it carries in lay use (ibid., 13). In their work incommensurability became crucial to address when the institution of the 'lay judge' was about to be introduced into civic courts. Consequently, their studies on the discrepancy between lay and expert use are not on how, say, witnesses or audience interpret the judges but on how lay judges participate in delivering justice. The semiotic differences are bound up with the material conditions that allow for legal deliberations to emerge as fair (on the materialities of law, see Pottage, 2012).

Leaving the courtroom challenges for the textual space of the Report on RoA, which I have already analysed as the circle of discredit, one can observe how multiple speech acts (legal definition, common use, institutional framework for audit, and Obokata's Appeal) are brought side by side as different truthful speeches. Veridiction works exactly in this opened-up space, where the performance and fate of these speech acts and their composed combination are rendered equivalent and subjected to evaluation. An excerpt may show the specific trope that allowed for the insertion of *akui* in the STAP case.

If we are to interpret "deliberate intent" as the intent to cause harm ... the Regulations would only apply when such strong intent is present, but it is evident this is contrary to the purpose of the Regulations which is to ensure the credibility of research papers, etc.

(Watanabe et al., 2014, 2)

A close examination of this argument shows that the Report's authors attribute to *akui* a *science-suitable* content, which omits harm as *irrelevant* to scientific regulations. Furthermore, they highlight how their interpretation is suitable to the STAP investigation, referring back to the use of the term 'intentionally' in Obokata's Appeal text:

> Therefore, we interpret "deliberate intent" or akui as ... "with knowledge or knowingly", and synonymous with "intentional". We note that the appellant uses the word "intentionally" herself, in her appeal when she states "I mistook the images. I did not intentionally submit different images" (Appeal document, p. 17).
>
> (Ibid., 2)

The above excerpts show how 'intentionality' was made obvious and measurable across the body of audit evidence or scientific knowledge gathered or generated by the STAP case investigation. The suggestive to-do list of quasi-legality was born out of the ambition to harmonize misconduct-related policies around a universal foundation that could accommodate cultural diversity and advance a template that would appear as fair and just. The latter was considered co-extensive with the proper scientific ethos. Within this normative idea of knowledge production as delivering truthful claims and thus undoubtedly *just claims*, scientists are expected to perform science-based detective work, becoming interrogators or judges in the courts of science. Drawing from the many misfortunes of the STAP case, I have tried to touch upon a number of shortcomings stemming from this framework.

Not surprisingly both MEXT and RIKEN reformulated the code of scientific conduct and related regulations in crucial points that diverge from quasi-legality (MEXT, 2014). Intentionality or intent as central categories are absent in these new guidelines. The only phrase that refers to them recognizes 'degrees of Intentions' that are to be approached objectively and not as a binary condition (ibid., 20). Notably, the new text dictates that allegations will not be received by internal offices but outside institutions will take the responsibility for organizing investigations in the future (ibid., 13). Furthermore, it dictates a careful disclosure of information in future instances of inquiries of fraud or misconduct, addressing for example the prevention information leaks to people outside the investigation (ibid., 15), controlled media coverage (ibid., 19), and accountability regarding suspicion (ibid., 20). Overall, the new guidelines address a series of organizational issues concerning the launch and management of an investigation that may be read as directly addressing the experience of August 2014. There is little trace of the legacy of the Tokyo and Lisbon proceedings.

The entanglement thesis: what remain the broader stakes of STAP?

In this chapter I have critically analysed the investigatory enterprise that claimed, in late 2014, to have resolved the STAP case as a case of research misconduct by the principal author, H. Obokata. The grounds of the investigation were constituted early on by an institutional attempt to control the circulating accounts generated by the initial flagging of STAP as a fraud case. Many layers of evidence collection and interpretative work can be identified throughout the deliberations of diverse expert bodies. The question of whether to maintain the distinction between auditing and reproducibility – as categories with distinct jurisdictions and systems of knowledge production – still remains. As I have shown, the documentation that was produced can be read as a hybrid of both audit processes and scientific experiments in a relationship of entangled parasitism, or simply *symbiosis* (*pace* Neyland): one fed onto the proliferation of the other's written accounts and vice versa, and both prosper. I have argued in support of a thesis of Barad-inspired entanglement: the sort of hybridity I point at is both the branching out of audit agendas into biological experiments and audit outcomes that literally build the spatial and virtual architecture of reproducibility attempts.

In the STAP case it is impossible to review the outcomes of auditing without reviewing the biology involved (mouse strain specificity) or reconfigured (genomic experiments), and in this chapter my aim was also to illustrate some of the painful, unjust, and authoritarian consequences of this entanglement. To recapitulate on my findings, in the first part I explored methods such as (1) interviews that inquire into scientific authorship and their audit-based and repetitive structure/scope, (2) the 'circle of discredit' as a potent tool in making unequal subjects-of-speech and proliferating the pejorative characterizations, and finally (3) the spatial re/organization of CDB laboratories and personnel throughout diverse experimental settings as an interrogation generating machinery.

In the second part, I built on the conclusions from this analysis and attempted to situate the STAP case within global desiderata regarding proper investigative methods. The origins of the entanglement thesis I suggest are found in the GSF's proposal to install scientists as expert-judges in the courts of technoscience. The quasi-legality of the suggested measures has been examined as a logic emerging in-between the orienting capacity of a to-do list for potential 'investigators' and the actual deployment of similar decisions, processes, and concepts by RIKEN. As the level of access afforded in this case has no precedent, one may wonder whether other misconduct investigations that proceed less publicly proceed with equal violence and vigour against the subjects and suspects of misconduct. This is debatable, but certainly a new area of inquiry has opened. For the field of STS, which pursues material/semiotic attentiveness to evidence-based practices and knowledge claims – and debates those *vis-à-vis* our values and institutions – the insights from the STAP case highlight how our methods and analysis tools may contribute as crucial f/actors in policy and implementation arenas that battle over what accounts as good practice and (mis)conduct in research. Precisely on the question of arenas of expert deliberations, the STAP case showcases the logic behind international desiderata and, in my opinion, stresses also the conceptual and practical void left by their shortcomings. This constitutes an open space where

STS research and intervention may support modes of responsible research and good practice that are empowering rather than coercive and unjust.

Finally, I attempted to show how the importance of the STAP case does not lie with the didactic morale for early career biologists (beware of Obokata's fate!), but in the unanticipated ways in which it has come to matter. It is not an anecdote or a ghost story of technoscience, but remains, in many aspects, an open case, at the very conjunction of a controversial epistemic field and governance proposal of research integrity. What I believe becomes apparent when insulating the content and arguments of the investigation from their media 'coverage' is that the STAP case *could have been otherwise* under different investigatory conditions – only if the whole design of its experimental attempts had proven more supportive to the people involved, less discriminatory in its argumentative constructions, and more attentive to questions inspired by a 'STAP biology'.

Notes

1 I refer to the first round of interviews, conducted between February and March 2014, and to the two letters submitted by the attorney of Obokata during the review of appeal.
2 This is also the main argument adopted by social networking platforms interested in post-publication peer-reviewing updates and information, e.g. the famous retraction watch and pubpeer.

References

Aizawa, S. (2015). Results of an attempt to reproduce the STAP phenomenon. *bioRxiv*, [online]. Available at: www.biorxiv.org/content/10.1101/028472v1 [Accessed 8 Mar. 2019].

Barad, K. (2010). Quantum entanglements and hauntological relations of inheritance: dis/continuities, spacetime enfoldings, and justice-to-come. *Derrida Today*, 3(2), pp. 240–268.

Foucault, M. (1995). *Discipline and Punish: The Birth of the Prison*, 2nd ed. New York: Vintage Books.

Foucault, M. ([1981] 2014). In: F. Brion and E. B. Harcourt, ed., *Wrong-Doing, Truth-Telling: The Function of Avowal in Justice*. Chicago, IL: University of Chicago Press.

Garforth, L. (2012). In/visibilities of research: seeing and knowing in STS. *Science, Technology, & Human Values*, 37(2), pp. 264–285.

Gayle, A. and Shimaoka, M. (2017). Public response to scientific misconduct: assessing changes in public sentiment toward the stimulus-triggered acquisition of pluripotency (STAP) cell case via Twitter. *JMIR Public Health and Surveillance*, [online], Volume 3(2). Available at: https://publichealth.jmir.org/2017/2/e21/ [Accessed 8 Mar. 2019].

Global Science Forum (2007). Best Practices for Ensuring Scientific Integrity and Preventing Misconduct. Organization for Economic Co-operation and Development, [online]. Available at: www.oecd.org/sti/sci-tech/40188303.pdf [Accessed 15 Feb. 2018].

Goodyear, D. (2016). The Stem-Cell Scandal. Rivalries, Intrigue, and Fraud in the World of Stem-Cell Research. *The New Yorker*, [online]. Available at: www. newyorker.com/magazine/2016/02/29/the-stem-cell-scandal [Accessed 15 Feb. 2018].

Ishii, S., Iwama, A., Koseki, H., Shinkai, Y., Taga, T. and Watanabe, J. (2014). Report on STAP Cell Research Paper Investigation. RIKEN, [online]. Available at: www3.riken.jp/stap/e/f1document1.pdf [Accessed 15 Feb. 2018].

Katsura, I., Igarashi, K., Ito, T., Omori, K., Kubota, T., Gokita, A. and Yonekawa, H. (2014). Report on STAP Cell Research Paper Investigation. RIKEN, [online]. Available at: www3.riken.jp/stap/e/c13document52.pdf [Accessed 8 Mar. 2019].

Kobayashi, T. (2016). Cases of research ethics education at graduate schools in Japan – Kyushu University's case. *Educational Alternatives*, 14, pp. 367–373, [online]. Available at: www.scientific-publications.net/fr/article/1001305/ [Accessed 15 Feb. 2018].

Lancaster, C. (2016). The acid test for biological science: STAP cells, trust, and replication. *Science and Engineering Ethics*, 22(1), pp. 147–167.

Martin, A. (2014). Japan Research Center Reboots After Scandal. *Wall Street Journal*, [online]. Available at: www.wsj.com/articles/japan-stem-cell-research-center-downsized-amid-scandal-1409112964 [Accessed 15 Feb. 2018].

Mayer, T. and Steneck, N. (2007). Final Report to ESF and ORI First World Conference on Research Integrity: Fostering Responsible Research. European Science Foundation (ESF) and US Office of Research Integrity (ORI, Department of Health and Human Services), [online]. Lisbon, Portugal. Available at: www. eurosfaire.prd.fr/7pc/doc/1207648775_official_final_conference_report.pdf [Accessed 15 Feb. 2018].

Meskus, M., Marelli, L. and D'Agostino, G. (2017). Research misconduct in the age of open science: the case of STAP stem cells. *Science as Culture*, 27(1), pp. 1–23.

MEXT (2014). Guidelines for Responding to Misconduct in Research. Ministry of Education, Culture, Sports, Science and Technology, [online]. Available at: www. mext.go.jp/a_menu/jinzai/fusei/__icsFiles/afieldfile/2015/07/13/1359618_01.pdf [Accessed 15 Feb. 2018].

MEXT (2015). White Paper on Science and Technology 2015 (Provisional Translation). Ministry of Education, Culture, Sports, Science and Technology, [online]. Available at: www.mext.go.jp/en/publication/whitepaper/title03/detail03/1372827. htm [Accessed 8 Mar. 2019].

Mizuno, M. and Nakamura, S. (2009). Mock trial and interpreters' choices of lexis. Issues involving lexicalization and re-lexicalization of the crime. In: M. Mahlberg, V. González-Díaz, and C. Smith, eds., Proceedings of the Corpus Linguistics Conference (CL2009) University of Liverpool, UK (20–23 July 2009), [online]. Available at: http://ucrel.lancs.ac.uk/publications/cl2009/#papers [Accessed 15 Feb. 2018].

Neyland, D. (2012). Parasitic accountability. *Organization*, 19(6), pp. 845–863.

Neyland, D. and Woolgar, S. (2002). Accountability in action?: the case of a database purchasing decision. *British Journal of Sociology*, 53(2), pp. 259–274.

Niwa, H. (2015). Investigation of the cellular reprogramming phenomenon referred to as stimulus-triggered acquisition of pluripotency (STAP). *bioRxiv*, [online]. Available at: www.biorxiv.org/content/early/2015/09/28/027730 [Accessed 15 Feb. 2018].

Normile, D. and Vogel, G. (2014). STAP Cells Succumb to Pressure. *Science | AAAS*, 344(6189), pp. 1215–1216.

Obokata, H., Sasai, Y., Niwa, H., Kadota, M., Andrabi, M., Takata, N., Tokoro, M., Terashita, Y., Yonemura, S., Vacanti, C. A. and Wakayama, T. (2014a). Bidirectional developmental potential in reprogrammed cells with acquired pluripotency. *Nature*, 505, pp. 676–680.

Obokata, H., Wakayama, T., Sasai, Y., Kojima, K., Vacanti, M. P., Niwa, H., Yamato, M. and Vacanti, C. A. (2014b). Stimulus-triggered fate conversion of somatic cells into pluripotency. *Nature*, 505, pp. 641–647.

Pels, P. J. (2000). The Trickster's Dilemma: Ethics and the Technologies of the Anthropological Self. In: M. Strathern, ed., *Audit Cultures. Anthropological Studies in Accountability, Ethics and the Academy*. London/New York: Routledge, pp. 135–172.

Pottage, A. (2012). The Materiality of What? *Journal of Law and Society*, 39(1), pp. 167–183.

Power, M. (1994). *The Audit Explosion*. London: Demos.

Shore, C. (2008). Audit culture and Illiberal governance: Universities and the politics of accountability. *Anthropological Theory*, 8, pp. 278–298.

Star, S. L. (1999). The ethnography of infrastructure. *American Behavioral Scientist*, 43(3), pp. 377–391.

Strathern, M. (1997). 'Improving ratings': audit in the British University system. *European Review*, 5(3), pp. 305–321.

Van Noorden, R. (2014). Japanese lab at centre of stem-cell scandal to be reformed. *Nature News Blog*, [online]. Available at: http://blogs.nature.com/news/2014/08/japanese-lab-at-centre-of-stem-cell-scandal-to-be-reformed.html [Accessed 15 Feb. 2018].

Vogel, G. and Normile, D. (2014). Nature reviewers not persuaded by initial STAP stem cell papers. *Science. AAAS*, [online]. Available at: http://sciencemag.org. [Accessed 15 Feb. 2018].

Watanabe, J., Iwama, A., Koseki, H., Shinkai, Y. and Taga, T. (2014). Report on Review of Appeal of STAP Cell Research Paper Investigation Results (Official Translation). RIKEN, [online]. Available at: www3.riken.jp/stap/e/p2document14.pdf [Accessed 15 Feb. 2018].

11 The ethics of embryonic stem cell research in Turkey

Exploring the moral reasoning of Muslim scholars

Ahmet Karakaya

Abstract

Human life is respected both in secular and religious terms. However, when a human life begins, when an embryo becomes a human and/or person, is not an easy question to answer. The very ambiguity of the biological process translates itself into a series of ambiguities in the moral area. This is especially true for a subject matter such as human embryonic stem cells that promise a wide array of possibilities concerning human health, while the related research poses grave moral questions. This paper, in the first part, discusses the moral status of the embryo from Islamic-religious perspectives. In the second part, based on previously conducted qualitative research, the paper reveals ethical positions of Muslim scholars living in Turkey and the normative religious principles on which Turkish Muslim scholars base their positions regarding embryonic stem cell research. Despite the heterogeneity of the normative religious principles presented, three normative religious principles, including benefits (*maslahah*) and harm (*mafsadah*), general rule (*'azimah*) and exemption (*rukhsah*), and inviolability (*ismah*), stand out amongst the others.

Introduction

In the wake of developments in biomedical research, such as organ transplantation, assisted reproductive technologies, bioenhancement, etc., our societies have increasingly become concerned with their related ethical problems. The very ambiguity of the biological processes involved in these developments translates into a series of dilemmas in the moral field. This is true, too, when the subject matter is human embryonic stem cells (hESC), which promises an array of possibilities concerning medical treatments while the research on them poses serious ethical questions.

While these bioethical discussions have arisen along with biotechnical achievements that have appeared and developed in the West, there is still a lack of long-term debates in Muslim-majority countries. This chapter reports Turkish Muslim scholars' positions and arguments towards embryonic stem cell research. Based on research previously conducted (Karakaya and Ilkiliç, 2016), I will elaborate on the normative religious principles upon which Turkish Muslim scholars base their positions regarding embryonic stem cell research.

The Turkish Muslim discursive landscape concerning hESC research will be described based on interviews with scholars in Turkey on the subject. These interviews were conducted in 2015 and covered both the moral status of the human embryo and hESC research undertaken by Ilhan Ilkiliç and me. Our aim was to establish the ethical assessments of hESC research according to Muslim scholars located in different parts of Turkey. It was the first interview-based study with Turkish Muslim scholars on their opinions and arguments concerning hESC research. Fifteen semi-structured face-to-face interviews were conducted with the respondents. The interviews were conducted with three groups of scholars including scholars from academia, a scholar from the Directorate of Religious Affairs, which is the senior official religious institution in Turkey, and independent scholars – scholars neither working in any universities nor in the Directorate of Religious Affairs. Respondents were interviewed in their native language, Turkish, which I translated into English. We then conducted a critical analysis of these interviews. Data obtained through these interviews have been classified into three main themes: the positions of Turkish Muslim scholars towards embryonic stem cell research, the stage at which the embryo achieves moral value, and the moral implications of ensoulment (Karakaya and Ilkiliç, 2016). The focus was on understanding the foundations of Turkish Muslim scholars' positions and arguments as well as their ethical reasoning.

The chapter thus adds a perspective to the dominant discussion about embryo research in the West that illustrates the reasoning of Muslim scholars in Turkey. The findings include the insight that because of the interpretive openness of the Islamic texts, there is a breadth of scholarly opinion in which some argue in favour of the research for therapeutic purposes.

Islamic background

I begin by introducing some essential concepts of Islam which I will refer to throughout the paper. The word 'Islam' is derived from the Arabic root *slm*, which includes meanings such as peace, unconditional surrender to God, and eternal salvation. Islamic teachings cover all fields of human activity, including spiritual and material, individual and social, educational and cultural, economic and political. The discipline (*'ilm*) in which the instructions that regulate the human activities are explained is called Islamic jurisprudence (*fiqh*). The subject matter of *fiqh* is the analysis of human actions, whether towards one another (human-human relations) or towards God (human-God relations). It engages the spiritual (*al-fiqh al-batin*) as well as civil aspects (*al-fiqh al-zahir*) of individual and community. Therefore, all human actions are subject to the regulation of *fiqh*. *Fiqh* in its historical process has been applied to the determination of a right action. Accordingly, as Reinhart claims: 'Islamic law and legal theory must be the true locus of the discussion of Islamic ethics' (1983, 186).

In order to seek a method of interpretation employed in the realm of *fiqh*, Muslim jurists developed a legal theory known as *usul al-fiqh*. Its main purpose, as Hallaq states, is 'to provide a juristic methodology and a herme-neutic that can be utilized in the formulation of rules on the basis of sources (*usul*)' (2009, 75). According to this legal theory, Islamic jurisprudential laws (*ahkam*, plural of *hukm*, meaning judgement, assessment) are known through their sources. The Qur'an and Sunnah are considered by all traditional schools of law as the primary sources of jurisprudence. It is believed that the Qur'an was revealed by God to the Prophet Muhammad (peace be upon him) and as such is the literal word of God, and thus the foundational source of laws.

The Sunnah as a whole comprises the sayings, actions, and precepts ex-pressed by the Prophet Muhammad. While the Qur'an contains the princi-ples (of the Islamic laws), the Sunnah is the form of these principles. Prophet Muhammad's life manifests how divine commandments (revealed as such in the Qur'an) can be applied to an individual Muslim life. Therefore, there are many verses in the Qur'an explaining the necessity of obedience to the Prophet Muhammad.[1]

Along with these primary sources follow secondary ones. When there is a case whose solution has not been explicitly stated in the Qur'an and Sun-nah, *ijtihad* (independent reasoning) is one of the most important secondary sources of Islamic jurisprudence. Other secondary sources include analogi-cal reasoning (*qiyas*), juristic preference (*istihsan*), and public interest (*istis-lah*) (Hallaq, 2009, 22–74).

However, it is important to highlight the following point: *ijtihad* is the Muslim scholars' endeavour to understand the Qur'anic statements and Prophetic Sunnah. In order to understand the Qur'anic verses Muslims interpret the Prophetic Sunnah, and Muslims' practices in the subsequent period, using their own minds. The dialogue between Prophet Muhammad and his companion Mu'adh shows not only the method of Islamic jurispru-dence but also the importance of independent reasoning. As conveyed in the Sunnah:

> The Messenger of Allah sent Mu'adh to Yemen, [the prophet asked Mu'adh,] How will you judge? Mu'adh said, "I will judge according to what is in Allah's Book." [The Prophet said] If it is not in Allah's Book? Mu'adh said: "Then with the Sunnah of the Messenger of Allah." [The Prophet said] If it is not in the Sunnah of the Messenger of Allah? Mu'adh said: "I will give in my view." [The Prophet said] All praise is due to Allah, the One Who made the messenger of the Messenger of Allah suitable.
>
> (at-Tirmidhi, 2007, Vol. 3, 185)

In order to solve emerging issues within the framework of Islamic jurispru-dential laws, *fiqh* applies these sources to the problems. hESC research has only developed in the last three decades and thus it is not possible to find laws directly relating to it in the Qur'an and the Sunnah. Therefore the rea-soning of contemporary Turkish Muslim jurists' *ijtihads* is key to under-standing Muslims' attitudes to this controversial issue.

Moral significance of early human embryos

Following the availability of human embryos in the laboratory, philosophers and theologians have addressed the question of their moral status in various ways. Debates on hESC research stem from different moral values attributed to the concept of 'human being'. Philosophers have long pondered whether being a member of *Homo sapiens* bequeaths exceptional moral status or not. If it does, the important question becomes that of defining the beginning of human life. Alternative positions do not consider conception or being a member of *Homo sapiens* as sufficient reasons for a biologically human entity to obtain the full moral value of a person (Macklin, 1983; Sullivan, 2001).

Philosophers who consider that embryos obtain moral value depending on certain physiological capacities make a distinction between being a human being and being a person. They might concede that the biological human being's life begins with conception, but only persons possess moral value. Among the philosophers who frame this distinction is John Locke (d. 1704):

> We must consider what person stands for; – which, I think, is a thinking intelligent being, that has reason and reflection, and can consider itself as itself, the same thinking thing, in different times and places; which it does only by that consciousness which is inseparable from thinking, and, as it seems to me, essential to it: it being impossible for anyone to perceive without perceiving that he does perceive. When we see, hear, smell, taste, feel, meditate, or will anything, we know that we do so.
>
> (Locke, 1999, 318–319)

Many hold that being genetically human does not suffice for the moral and legal status of personhood. Some go further and define the necessary properties a human individual must possess to be a person. Michael Tooley, for instance, defines properties such as the capacity to envisage a future for oneself, and to have desires about one's future state, the capacity to have a concept of self, being a self, self-consciousness, the capacity for self-consciousness which gives the organism a right to life (Tooley, 1972). Other philosophers have stressed the tension between the continuity of the embryonic development and the need to assign status in order to regulate human reproductive medicine (Warnock, 1983). The consequences of the distinction between being human as a species category and personhood as a moral legal status have led many medical ethicists to determine the point in time or developmental stage at which an embryo achieves moral value. Using biological and medical knowledge, they have suggested demarcation points such as fertilization, implantation, formation of the primitive streak, sentience, quickening, physical formation, viability, and ensoulment. However, all these demarcation points have been subject to criticism, without a unanimous conclusion being reached (Aksoy, 1998; Cohen, 2007). After a brief introduction to the wider context of the debate, I will show that some of these positions are also asserted by Turkish Muslim scholars.

Embryonic stem cell research in Turkey

Islam is not only a faith but also a legal and moral system aiming at organizing the whole of life. Accordingly, Islamic teachings cover all human activity including its spiritual and material dimensions (al-Faruqi, 1992). The first reaction to Islam's claim to administer all fields of life in the Ottoman Empire came during the Tanzimat period in the nineteenth century. Shortly after the collapse of the Ottoman Empire and the abolition of the caliphate in 1924, the newly established Turkish Republic, unlike its predecessor, adopted secularism as the official state ideology. Turkey, now, as indicated clearly in its constitution is a secular state. However, the structure of religious institutions is not consistent with the principle of secularism as understood in the West. For instance, although the application of secularism in the West might allow the existence of autonomous religious authority, it is not the case in Turkey. The Presidency of Religious Affairs is the senior religious institution in Turkey and it is under the legal control of the Turkish government. For a foreign audience, the implementation of secularism in Turkey can appear inconsistent (Adanali, 2008).

In order to discuss the Islamic-religious debates on embryonic stem cell research in Turkey, we need to rethink the principles of secularism, its emergence in the West, and applications in non-Western countries. We also need to take into account the very basic Islamic norms that regulate a Muslim's life. Therefore, due to the Islamic worldview and current Turkish model of secularism, I will not draw a strict distinction between secular and Islamic-religious perspectives in this chapter. I believe that although state institutions are based on secular juridical norms, Islamic ethical norms play a significant role in defining public attitudes towards controversial issues in modern Turkey – especially when backed by the opinions of domestic Muslim scholars.

Before elaborating on the issue, it is important to outline the attitude of the Presidency of Religious Affairs and the current legal regulations on stem cell research in Turkey briefly to explain the context in which scholars shape their views. The High Council of Religious Affairs is the highest decision-making and advisory body on religious issues within the Presidency of Religious Affairs. According to the advisory opinion letter issued by the High Council in 2006, the source of stem cells is of great importance in the discussion of the legality of stem cell research. The use of adult stem cells from tissues and organs such as bone marrow and cord blood is seen as similar to organ transplantation. The Council does not see any drawback in transplanting such stem cells under circumstances in which organ transplantation is permissible.[2] Concerning hESC research, the High Council notes that the idea that a human embryo is but a cluster of cells reflects a pure materialist understanding and contradicts Islamic thought. The embryo is the material source of the human body and should be treated with respect and honour. However, the High Council does not adopt an exclusionary view against hESC research. It states that scientists should intensify their studies on adult stem cells and use them to find medical treatments, but if scientists cannot make therapeutic use of adult stem cells in the same way in which hESC can be used, the excess embryos in IVF units can be used for hESC research. It is important to note, though, that hESC might be allowable only for medical treatment, if there is no other way and not for commercial ends (High Council of Religious Affairs, 2017).

Apart from these opinions of the High Council, there is not any regulation concerning hESC research in law in Turkey. In 2005, the Turkish Ministry of Health issued a circular letter, which is still valid for scientific practice, to define the legal framework of hESC research in Turkey (Turkish Ministry of Health General Directorate of Treatment Services, 2005). This circular letter states that various studies had been conducted on the use of stem cells in medicine both in Turkey and elsewhere in the world. Even though the results of these studies held great promise regarding the use of stem cells for the treatment of many diseases, they also triggered different debates depending on the sources of the stem cells used. According to this circular letter,

> The Ministry of Health has been working on legal adjustments for embryonic stem cell research that need to be realized in accordance with the requirements of modern science and public consciousness; therefore, embryonic stem cell research should not be conducted until the Ministry of Health finalizes the regulations in this area.

Even though the Ministry underlined that the research was not principally banned, a legal regulation has not hitherto been issued. In a second circular letter (Turkish Ministry of Health General Directorate of Treatment Services, 2006), the Turkish Ministry of Health allowed adult stem cell research to be conducted in Turkey and defined the legal and regulatory framework for adult stem cell research (Karakaya and Ilkiliç, 2016).

Islamic-religious perspectives: moral status of the embryo and embryonic stem cell research

Ethical debates on hESC research have centred almost exclusively on the question of the moral status of the early human embryo, especially in Western societies (critically noted by Bender et al., 2005; Hauskeller, 2005). My initial motivation was the wish to see how the question of the moral status of early human embryos played a part in Muslim scholars' moral reasoning. That is why I became interested in exploring the arguments held by Turkish Muslims scholars on hESC research.

Before elaborating on these arguments, it is important to outline the Turkish Muslims' opinions on the moral status of early human embryos. Based on a literature analysis and the aforementioned interviews, I found

that Turkish Muslim scholars have addressed the moral status of the embryo from different perspectives and have pointed out three key demarcation points including conception, implantation, and ensoulment. Some
scholars consider conception as the beginning of a human being's life from
a moral perspective, some argue that embryos are morally valuable from the
first moment of implantation, and a third group of scholars share the view
that ensoulment marks the moment when the embryo becomes a human
being. I will now present all three positions, respectively.

Conception

Our interview findings show that the majority of the Turkish Muslim
scholars consider conception as the beginning of an individual human being's life from a moral perspective. Their basic argument is that the embryo is a unique individual whose genetic code has already been defined.
Human life is a process that should not be interfered with by arbitrary
divisions; the human embryo is one of the stages of the human developmental process, and the embryo is a potential human being that would
naturally develop to become a complete human. The majority of Turkish
Muslim scholars also argue that while the human embryo has the potential
of becoming a human it is not actually a human being yet. Therefore, it
has a special status but this neither entails that the human embryo does
not have any value nor that it has all the moral rights of a human being.
According to them, the embryo is the core of the human being and thus it
should be protected from the first moment of conception irrespective of its
developmental stages:

> Human life begins at the moment of conception and unfolds until the
> moment of birth. The assessment of the consequences arising from an
> intervention into the development of this potential human in the moth-
> er's womb varies according to the stage of its progress. The sanction of
> intervention into the embryo increases as time goes on.
>
> (Interview, 20 February 2015)

According to this view, intervention into the embryo can only be justified in
crucial cases, and the limits of such interference are set by a group of Muslim scholars. Crucial cases may include life-threatening situations for the
mother, defined as *zarurah* (a very urging need), and cannot be established
by individual scholars themselves.

Implantation

A view maintained by other Turkish Muslim scholars is that the life of the
embryo begins with the implantation of the fertilized egg in the mother's
womb. Their argument is that there can be no survival of a human embryo
outside the womb, and therefore, an embryo which is not implanted in the

uterus cannot survive. Hence human life begins with the embryo's implantation in the womb. That viewpoint entails that a main condition of being a human is the environment, and emphasizes that it is wrong to reduce human beings to mere genetic codes:

> When a fertilized egg is implanted in the mother's womb, it begins to be nourished and in the end, a human being is born. Any intervention with this embryo should not be allowed. But since an embryo that is not implanted in the uterus is not being nourished and the process does not lead to a birth, you cannot consider this embryo as a human being and it can be destroyed.
>
> (Interview, 17 January 2015)

Ensoulment

Ensoulment has been one of the most important arguments regarding any medical intervention affecting the embryo, including embryonic stem cell research and abortion. According to some Turkish Muslim scholars, although biological life begins with conception, the organism becomes a human being with ensoulment. The termination of embryo development is forbidden after ensoulment has occurred.

Although ensoulment is not explicitly described in the Qur'an,[3] the concept of the soul is referred to in several different verses of the Qur'an in various contexts. The term 'soul' is referred to as a revelation in al-Nahl 16/2, al-Ghafir 40/15, and al-Shura 42/52 – and to the angel Gabriel in Maryam 19/17, and al-Shu'ara 26/193. Further references are to the soul as a special divine support in al-Baqarah 2/253, al-Ma'idah 5/110, and as an essence that makes a human being a person in al-Sajdah 32/9, al-Hijr 15/29, and Sad 38/72. Among these verses two will be presented in more detail to illustrate the Qur'anic view of the ensoulment:

The first verse is:

> He Who has made everything which He has created most good: He began the creation of man with (nothing more than) clay, and made his progeny from a quintessence of the nature of a fluid despised. But He fashioned him in due proportion, and breathed into him something of His spirit. And He gave you (the faculties of) hearing and sight and feeling (and understanding): little thanks do ye give!
>
> (Holy Qur'an; al-Sajdah 32: 7–9)

The second is:

> Behold, thy Lord said to the angels: "I am about to create man from clay: When I have fashioned him (in due proportion) and breathed into him of My spirit, fall ye down in obeisance unto him".
>
> (Holy Qur'an; as-S'ad 38: 71–72)

Since the Qur'an does not specify an exact time of ensoulment, some scholars have been trying to establish this point in time based on some hadiths.[4] Although a survey of the classical literature regarding the moment of ensoulment shows that the majority of the scholars believe this event to occur on the hundred and twentieth day of pregnancy (Ilkiliç and Ertin, 2010), a common view among contemporary Turkish Muslim scholars is the fortieth day (Karakaya and Ilkiliç, 2016). Although the time of ensoulment seems to be determined through these hadiths, this does not mean that there is no difference in opinion among individual scholars.

However, some scholars argue that the time of ensoulment cannot be determined through these hadiths. Hayrettin Karaman, a prominent Islamic law scholar, states that since there are different narrations of these hadiths, the determination of an exact time for the ensoulment based on these hadiths is not justifiable. According to him, these hadiths do not give any information regarding the exact day of ensoulment (Karaman, 2012).

Two factors are important to stress as reasons for the differences in opinion among Turkish scholars regarding the time and nature of ensoulment. One is the lack of an ultimate centralized religious authority in Turkey, in contrast to many Christian countries; the other is the ambiguities in the relevant verses which allow for different interpretations. The resulting understandings of the time of ensoulment can be summarized as follows:

i   Although the life of an organism begins at conception, it becomes a human being with ensoulment.
ii The related verses and hadiths do not give any definitive information regarding the day of ensoulment. Their purpose is to provide lessons to understand God's existence, unity, and will. These verses and hadiths are intended to explain the power of God and thus do not define a day of ensoulment. Therefore, ensoulment cannot be used as a religious argument regarding any medical intervention affecting the embryo.
iii Although the common view among contemporary Turkish scholars is that ensoulment happens on the fortieth day of gestation according to the Qur'an, it is not appropriate to assign a point in time based on these hadiths. Ensoulment may happen at any time after the first moment of conception.

Positions and arguments of Turkish Muslim scholars towards embryonic stem cell research

There has been a variety of approaches and an ensuing number of debates about using human embryos in hESC research among Turkish scholars. Their positions and arguments are not homogenous. Some scholars argue that embryos should not be used at all; others favour using embryos in stem cell research. I will explain the positions of Turkish Muslim scholars distinguishing arguments that (a) leave the option of hESC research open and others (b) that oppose it.

a   Scholars who defend the use of embryos in stem cell research take one of two positions: (i) embryos can be used only before implantation, and (ii) embryos can be used before ensoulment.

The number of scholars in the first group is not large.[5] According to them, given that the embryo which is not implanted in the uterus cannot survive, it cannot be considered a human being, and therefore can be used for stem cell research. Likewise, using excess embryos in IVF units is permissible in this research, unless they are implanted in a woman's womb. Hayrettin Karaman states:

> Despite some objections, assisted reproductive technology is well accepted by Muslim scholars, provided that both sperm and ovum are taken from married wife and husband. But the problem of fertilizing more than one ovum and creating excess embryos, then destroying some of these embryos and implanting the healthiest ones into the uterus remains the important part of the issue. Also, questions arise such as whether the excess embryos which are not implanted into the uterus should be considered as real embryos and if it is possible to destroy those embryos once created. My answer to these questions is as follows: When a fertilized egg is implanted into the uterus, it begins to be nourished and to develop there, and finally a human will be born, unless the process is being interfered with. Therefore this is an embryo, and destroying this embryo is not allowed. Since an embryo which has not been implanted cannot develop and be born, it is not a human being and can be destroyed.
>
> (Karaman, 2012, 75)

According to scholars in the second group, the definition of death from an Islamic concept is the irreversible complete loss of function in three bodily systems – namely, the cardiovascular, the respiratory, and the neural system. From this point of view, the beginning of human life is the point at which any one of these three systems begins to form. Hence it is the formation of the cardiovascular system, when the embryo's heart begins to beat. In the case of hESC research, considering the medical benefit to be gained from it, this group of scholars maintains that Islam not only permits such research but advocates it.

> In my opinion, anything that is produced in the laboratory may also be destroyed by humans. Because if you are able to fertilize sperm and egg in the IVF units, that's because God allows you to do so. Therefore, treatments which are made in the area permitted by God cannot be judged as being contrary to God's creation. Islam

does not only allow this research but encourages it. In this sense, not only excess embryos stored in IVF units but also all laboratory-produced embryos can be used in stem cell research.

(Interview, 30 October 2014)

According to this view, the main reason for the prostration of angels before the human being which God mentions in the Qur'an is ensoulment (Holy Qur'an; Al-Hicr 15: 28–29). As ensoulment, according to them, is a process in which intellect (*aql*), will (*iradah*), and conscience (*wijdan*) emerged in a human being, it is not an instantaneous event; rather, it is an ongoing process which starts at the beginning of the formation of the cardiovascular system in the embryo and ends when intellect, will, and conscience are formed in the human being. The liability of the human being to God's commandments thus begins after the ensoulment takes place.

b   Scholars opposing the use of embryos mainly consider conception as the beginning of a human being's life and therefore the point where morality applies.

> Evaluating the beginning of human life and the issue of the use of excess embryos in IVF units, this group of scholars suggests that the time of the real and logical beginning of the single person as an individual is when ovum and sperm, which originate as two separate and independent entities, fuse and thereby activate a process to form a complete human being. Therefore, they do not take a positive view of using embryos to obtain hESCs. Regarding excess embryos created in IVF units and their use in hESC research, Saim Yeprem, a former member of the High Council of Religious Affairs states: In order to remove such a drawback, no more ova than necessary should be inseminated if possible. In other words, only the ova that are needed should be inseminated. Otherwise, destruction of the remaining inseminated ova would be problematic from religious perspective.On the other hand, in cases where there is a necessity to produce more embryos than necessary due to medical and technical inevitability, the process should be kept at minimum as much as possible and the remaining should be allotted to stem cell research instead of destructing them.

(Yeprem, 2007, 45)

This quotation contains a typical demand among Turkish Muslim scholars who oppose using human embryos in stem cell research. They argue that the foremost criticism towards hESC research should be directed at the excess embryos created in IVF units. They urge that no more ova than necessary should be fertilized if possible and support the idea of creating just one embryo and implanting it into the mother's womb. This is not current practice in IVF clinics in Turkey. These scholars assume that human life begins with conception and reject the use of embryos resulting from *in vivo-* or *in vitro-*fertilization in scientific research. Yet, in cases where it is necessary to produce more embryos than will be implanted for medical and technical reasons, the process should be kept to a minimum and the remaining cells or zygotes could then be allocated to hESC research instead of being destroyed. Moreover, these scholars argue that embryos which are obtained through miscarriages can be used in stem cell research.

Discussion

My research shows that although the majority of Turkish Muslim scholars favour the protection of embryos from the first moment of conception, they still support using human embryos in hESC research. It may seem contradictory to find scholars who support the use of embryos for hESCR while at the same time indicating that embryos achieve moral value at conception. This attitude of Turkish Muslim scholars motivates me to pursue normative religious norms upon which these scholars base their arguments, and their moral reasoning as well. A result of my analysis is that Turkish Muslim scholars use several religious norms to evaluate the moral aspect of hESC research. The most prominent of them are benefits (*maslahah*) and harm (*mafsadah*), general rule (*'azimah*) and exemption (*rukhsah*), and the inviolability of human life (*ismah*). I will introduce them briefly.

Benefits (*maslahah*) and harm (*mafsadah*) in Turkish Muslim reasoning

Linguistically *maslahah* means the utmost righteousness and goodness (*salah*). In Islamic legal theory, *maslahah* is used to promote benefits to the public or to individuals, and to prevent social evil or corruption. *Mafsadah* is the antonym of *maslahah*, meaning that it is an attribute of the act whereby corruption or harm happens to the public or to individuals (Ibn Ashur, 2006, 96–99). According to Al-Ghazali (2007, 120), one of the most prominent and influential jurists, theologians, and Sufis of the twelfth century, *maslahah* is that consideration which secures a benefit or prevents harm and is, at the same time, harmonious with the aims and objectives of Islamic law (*Shari'a*). These objectives consist of protecting the five essential values – namely religion, life, intellect, lineage, and property. According to Al-Ghazali, any measure which secures these values falls within the scope of *maslahah*, and anything which contravenes them is *mafsadah*; preventing the latter is also *maslahah* (ibid.). Considering this conceptual framework, the idea is widely accepted among Turkish Muslim scholars that Islam encourages and supports everything that benefits individuals and society, provided that the benefit (*maslahah*) outweighs the harm (*mafsadah*). According to some Muslim scholars,

Islam is the religion of weighing the *maslahah* and the *mafsadah*. Concerning a complex jurisprudential issue like embryonic stem cell research, we (scholars) usually judge the actions, which include both aspects of *maslahah* and *mafsadah*, according to the predominant aspect. If the *maslahah* outweighs the harm, then the action is desired. If, on the contrary, *mafsadah* outweighs the benefit, then the action is to be avoided.

(Interview, 3 May 2015)

According to the Turkish Muslim scholars we have interviewed, the complex character of hESC research and the ambiguity of the biological process obstruct the formulation of the *maslahah* and *mafsadah* of the research. Opponents, especially those who think that the life of the embryo begins with conception and that it is the foremost aim and objective of the *Shari'a* to consider the life of a human being, consider the *mafsadah* of the research to be the dominant concern – embryos should not be used in research. But the majority of our Turkish Muslim scholars hold the view that the medical benefit of stem cell research is a great *maslahah* for society and for individuals, prevailing over *mafsadah*. According to them, human embryos can be used in research (Karakaya and Ilkiliç, 2016).

General rule (*'azimah*) and exemption (*rukhsah*)

The lexical meaning of *rukhsah* is softening. According to Islamic law, the Lawgiver may indicate that one law is to be considered as an obligation imposed initially as a general rule (*'azimah*). This may be followed by another rule that is an exemption (*rukhsah*) from the general rule. According to Al-Shatibi, a prominent Andalusian scholar of the fourteenth century:

> *'azimah* is based upon an obligation creating a principle that is universal. Because it is absolute and general in its application to all subjects. *Rukhsah* on the other hand is based upon a particular rule that applies to some of the subjects who have an excuse, such as special circumstances or crisis in their life. It is not available under all circumstances and at all times, nor does it apply to everyone. It is like a temporary obstacle affecting the universal.

(al-Shatıbi, 2003, 304)

For instance, eating pork is prohibited as a general rule, but according to *rukhsah*, a Muslim who is under extreme pressure such as famine may eat pork if it saves that Muslim from dying of hunger (Holy Qur'an; al-An'am 6: 145). According to some scholars who defend the use of embryos in stem cell research, treating diseases is seen as one of the necessities that needs to be solved. Therefore if the research on hESC is successful in treating diseases then doctors may use and even procure embryos for therapeutic purposes.

The Islamic concept of inviolability (*Ismah*)

Another normative religious principle of those who accept the use of human embryos in stem cell research is based on the understanding of the inviolability of the human body, an understanding which differs from one Islamic school of thought to another. According to this view, there is no obligation for a Muslim to obey the *Hanafi* school of thought, a school followed by the majority of Turkish people, in all areas of life. Considering the medical benefits to be gained from stem cell research, embryos may be used in this kind of research following the doctrines of the *Shafii*, which allows the use of any part of the human body in extreme situations.[6]

> Human beings are honourable and therefore it is prohibited to profane their bodies. The utilization of any part of the human being, including any organ or tissue, including hair, by any means, is unacceptable. However, in the Shafii school of thought, in accordance with a public interest, one can use any part of the human body. For instance, a Muslim who is helpless in the face of death can eat a piece of a human body if and only if this saves him/her from dying.
>
> (Interview, 5 April 2015)

In addition to these religious norms, having a healthy body is also regarded as a religious obligation by Turkish Muslim scholars. Accordingly, health is considered one of the greatest of God's blessings to humans. It can be considered the greatest blessing God gave to human beings after faith – as narrated by the Prophet Muhammad: 'Ask Allah for pardon and *Al-Afiyah* [good health] for verily, none has been given anything better than *Al-Afiyah*' (at-Tirmidhi, Vol. 6, 2007, 265). Muslims consider their body as an entrusted gift from God, and they believe they are not the owners of their bodies. This imposes certain responsibilities upon Muslims, including seeking treatment for diseases and keeping the body healthy. Therefore, the majority of scholars consider seeking treatment and engaging in disease-preventing precautions, to be religious obligations. Given its central meaning and importance, some scholars consider contributing to health the most important reason (*maslahah*) that can legitimize the use of human embryos in stem cell research. But this does not mean that this opinion is accepted by all Muslim scholars.

Conclusion

There is much literature emphasizing the decisive role that the status of the human embryo plays with regard to the ethics of stem cell research. Turkish Muslim scholars have addressed the moral status of the embryo from different perspectives and have identified three key demarcation points: conception, implantation, and ensoulment. However, although the majority of these scholars consider conception as the beginning of an individual human being's life from a moral perspective, they still favour using human embryos in stem cell research.

My research also shows that there are several normative religious principles used by Turkish Muslim scholars which counterbalance the moral status of the early human embryo. These principles, which Turkish Muslim scholars base their positions on, are benefits (*maslahah*) and harm (*mafsadah*), general rule (*'azimah*) and exemption (*rukhsah*), and inviolability (*ismah*).

Although these normative religious principles have led to the formation of different positions in the ethical assessment of embryonic stem cell research, the dominant position towards this kind of research is favourable. The current global discursive template on how to assess the views of non-Western societies on hESC research is dominated by the ethical and moral stances proposed within their own moral frameworks. Therefore, it would seem helpful to relinquish the dominant discourse on concepts of personhood and the moral status of the human embryo in order to understand the authentic view of Turkish Muslim scholars on hESC research.

When working through the arguments brought forth by Turkish Muslim scholars regarding the permissibility of hESC with the dominant Western perspective, we find that their debates thematize and reflect on the moral status of the embryo and the concept of personhood on the one hand – yet on the other hand, they do so in specific ways and in relation to other religious norms and obligations such as a healthy body and advances in medicine. The hierarchy of these norms is the issue, and in that the majority Turkish Muslim view differs from prominent prohibitive positions taken by certain Western churches and philosophers.

Acknowledgements

The author wishes to express his gratitude to the scholars who gave their time to participate in this study. He also thanks Christine Hauskeller, Hakan Ertin, Ilhan Ilkiliç, and Rainer Brömer for commenting on earlier versions of this chapter. The author further wants to thank Arne Manzeschke and Anja Pichl for inviting me to the TTN Summer School which opened up to me a new career trajectory.

Notes

1 Holy Qur'an; Al-Maide 5:92, An-Nisa 4:80, Al-i Imran 3:31, Al-Hasr 59:7, An-Nisa 4:65, Al-Ahzab 33:36, An-Nur 24:63.
2 For the Turkish regulation of organ transplantation please see: www2.diyanet. gov.tr/dinisleriyuksekkurulu/Sayfalar/OrganNakli.aspx and www.saglik.gov.tr/ TR,10528/organ-ve-doku-nakli-hizmetleri-yonetmeligi.html.
3 All quotations from the Qur'an are from Yusuf Ali's 1996 translation.
4 The hadith related to ensoulment is:

> The creation of any one of you is put together in his mother's womb for forty days, then, he is during that (period) an *'Alaqah* for a similar period. Then he

becomes a *Mudghah* for a similar period. Then Allah sends to him an angel who breathes the soul into him, and is enjoined to write down four things: His provision, his lifespan, his deeds and his misery or happiness. By the One besides Whom none has the right to be worshiped!

However, there are different narrations of this hadith stating the ensoulment takes place on the fortieth, forty-second, forty-fifth, eightieth, and hundred-and-twentieth day of pregnancy. For the details of these hadiths see al-Hajjaj (Vol. 7, 2007). Therefore, ensoulment is assumed to occur between the fortieth and the hundred-and-twentieth day of pregnancy in the classical literature.

5 Two scholars of fifteen.
6 In Sunni Islam, there are four doctrinal legal schools. These schools are: the Hanafi, Maliki, Shafi'i, and Hanbali, named after the four master-jurists who are assumed to be their founders. For detailed information see Hallaq (2009, 22–74).

References

Adanali, A. H. (2008). The presidency of religious affairs and the principle of secularism in turkey. *The Muslim World*, 98(2–3), pp. 228–241.

Aksoy, S. (1998). Ethical aspect of decision-making with regard to handicapped prenates and newborns. PhD thesis, University of Manchester.

Bender, W. et al. (2005). *Crossing Borders: Cultural, Religious and Political Differences Concerning Stem Cell Research, A Global Approach*. Münster: Agenda Verlag.

Cohen, C. B. (2007). *Renewing the Stuff of Life, Stem Cells, Ethics, and Public Policy*. Oxford: Oxford University Press.

al-Faruqi, I. R. (1992). *Al Tawhid, Its Implications for Thought and Life*. 2nd ed. Virginia: International Institute of Islamic Thought.

al-Ghazali, Ebu H. M. (2007). *Mustasfa (in Turkish)* (trans. Yunus Apaydın). Istanbul: Klasik Yayınları.

al-Hajjaj, H. M. (2007). *Sahih Muslim*, Vol. 7 (trans. Nasiruddin al-Khattab). Riyadh: Maktaba Dar-us-Salam.

Hallaq, W. (2009). *Shari'a: Theory, Practice, Transformations*. Cambridge: Cambridge University Press.

Hauskeller, C. (2005). Introduction. In: W. Bender, C. Hauskeller and A. Manzei, eds., *Crossing Borders: Cultural, Religious and Political Differences Concerning Stem Cell Research, A Global Approach*. Münster: Agenda Verlag, pp. 9–24.

High Council of Religious Affairs. (2017). *Evaluation of New Medical Practices such as In Vitro Fertilisation and Stem Cell Research Discussed in Today's Medical World in Terms of Islamic Religion*, [online]. Available at: www2.diyanet.gov.tr/dinisleriyuksekkurulu/Sayfalar/Tupbebek1023–5894.aspx [Accessed 14 July 2017].

Ibn Ashur, M. T. (2006). *Treatise on Maqasid Al-Shari'ah*. King's Lynn: International Institute of Islamic Thought.

Ilkiliç, I. and Ertin, H. (2010). Ethical aspects of human embryonic stem cell research in the Islamic world: positions and reflections. *Stem Cell Reviews and Reports*, 6(2), pp. 151–161.

Karakaya, A. and Ilkiliç, I. (2016). Ethical assessment of human embryonic stem cell research according to Turkish Muslim scholars: first critical analysis and some reflections. *Stem Cell Reviews and Reports*, 12(4), pp. 385–393.

Karaman, H. (2012). *Islam in Our Life (in Turkish)*. 5th ed. Istanbul: Iz Yayincilik.

Locke, J. (1999). *Essay Concerning Human Understanding*. University Park, PA: Pennsylvania State University.

Macklin, R. (1983). Personhood in the bioethics literature. *The Milbank Memorial Fund Quarterly. Health and Society*, 61(1), pp. 35–57.

Reinhart, A. K. (1983). Islamic Law as Islamic Ethics. *The Journal of Religious Ethics*, 11(2), pp. 186–203.

al-Shatıbi, E. I. (2003). *Muvafakat (in Turkish)* (trans. Mehmet Erdoğan). Istanbul: İz Yayıncılık.

Sullivan, D. M. (2001). A thirty-year perspective on personhood: how has the debate changed? *Ethics and Medicine*, 17(3), pp. 177–186.

at-Tirmidhi, M. E. (2007). *Jami' At-Tirmidhi*, Vol. 3 (trans. Abu Khaliyl). Riyadh: Maktaba Dar-us-Salam.

at-Tirmidhi, M. E. (2007). *Jami' At-Tirmidhi*, Vol. 6 (trans. Abu Khaliyl). Riyadh: Maktaba Dar-us-Salam.

Tooley, M. (1972). Abortion and Infanticide. *Philosophy and Public Affairs*, 2(1), pp. 37–65.

Turkish Ministry of Health General Directorate of Treatment Services. (2005). Circular Note of Turkish Ministry of Health on Embryonic Stem Cell Research, Circular Number 2005/141, [online]. Available at: www.ttb.org.tr/mevzuat/index.php?option=com_content&view=article&id=347:embron-k-hre-arairmalari-hakkinda-saik-bakanli-genelges&catid=3:tebligenelge&Itemid=35.   [Accessed 14 July 2017].

Turkish Ministry of Health General Directorate of Treatment Services. (2006). Circular Note of Turkish Ministry of Health on Stem Cell Studies, Circular Number 2006/51, [online]. Available at: www.ttb.org.tr/mevzuat/index.php?option=com_content&task=view&id=387&Itemid=35 [Accessed 14 July 2017].

Warnock, M. (1983). In Vitro Fertilization: The ethical issues (II). *The Philosophical Quarterly*, 33(132), pp. 238–249.

Yeprem, S. (2007). Current assisted reproduction treatment practices from an Islamic perspective. *Ethics, Law and Moral Philosophy of Reproductive Biomedicine*, 14(1), pp. 44–47.

Yusuf Ali, A. (1996). *The meanings of The Holy Qur'an*, [online]. Available at: www.wright-house.com/religions/islam/Quran.html [Accessed 9 Oct. 2017].

12 Balancing social justice and risk management in the governance of gene drive technology

Lessons from stem cell research

Achim Rosemann

Abstract

This chapter examines challenges related to the emergence of international governance frameworks for stem cell and gene drive research, from a social justice perspective. The chapter combines insights from current research on the development of new regulatory models for gene drive technology with findings from a previous study on the governance of stem cell medicine. The development of regulatory frameworks for both of these technology fields is still at a relatively early stage, and the formation of standardized international regulations has been contested. At the heart of these conflicts is a tension between the management of technology risks and the realization of social justice, conceptualized not only in terms of access to the potential benefits of these two technology fields but also in terms of access to new business and innovation opportunities by researchers and technology producers, especially in the context of low- and middle-income countries. The chapter explores these processes and suggests that in the light of the differing types of risks of gene drive and stem cell research, very different dilemmas and challenges for the governance of these two technology fields arise. This situation also raises important questions with regard to the role, theorization, and limits of the social justice concept in processes of technology governance – especially since the realization of social justice for some can result in increased risk and new forms of injustice for others.

Introduction

Stem cell and gene drive research are two emerging areas of biotechnology that have generated extensive controversies and challenges for responsible forms of international governance. While the technical aspects, methodologies, and applications of these two technology fields differ in important respects, what they share is that the development of internationally harmonized governance frameworks is contested and difficult to implement, especially in the context of low- and middle-income countries. This chapter shows that at the heart of these regulatory conflicts lies a tension between the management of technology risks and the realization of social justice, defined not only in terms of access to potential benefits by technology users, but also in terms of access to new business and innovation opportunities by researchers and technology producers. This tension is particularly pronounced in the context of developing and economic threshold countries, but as recent developments in the stem cell field show, similar developments also take place in the context of high-income countries.

Gene drives are a form of genetic engineering that allow for the spreading of gene modifications in species at a rapid pace, and potentially at the ecosystem-wide level. Gene drive applications are associated with scientific, agricultural, and health benefits – but also with numerous uncertainties, environmental risks, irresponsible commercial applications, and issues related to biosecurity, such as the use of gene drives for military purposes or bioterrorism. At present, countries are in the early stages of examining regulatory possibilities for gene drive applications. While there is a consensus that international governance strategies are needed, current multi-lateral fora focus primarily on the development of (highly expensive) science-based strategies to manage risks and possible adverse effects. One example is the Committee for Gene Drive Research in Non-Human Organisms, from the US National Academy of the Sciences, which has published an influential report on gene drive governance (National Academies, 2016). Participants in this and other international fora have overwhelmingly been from high-income countries, which indicates an over-representation of the interests, outlook, and expectations of stakeholders in highly developed, democratic societies. This poses important questions with regard to the feasibility of developing regulatory standards for low- to middle-income countries.

A discussion of the ways in which social, economic, regulatory, scientific, and healthcare inequalities and contexts in low- to middle-income countries influence the acceptance, adoption, and implementation of 'international' regulatory instruments for gene drive applications has hitherto mostly remained absent. This applies not only to gene drive technology but also to other emerging areas in the life and health sciences. This is problematic for a variety of reasons. As Sleeboom-Faulkner and colleagues have shown in a study of regulatory developments for clinical stem cell research in several Asian societies, stakeholders in scientifically advancing, low- to middle-income countries often see the adoption of 'international' technology standards as a hindrance to the realization of domestic innovation and corresponding economic profits. Many governments, scientists, and local corporations do not possess the financial, technological, and infrastructural resources to ensure consistent compliance with international regulatory demands, as developed in high-income countries (Sleeboom-Faulkner et al., 2016). At the same time, in many biotechnology fields the development and implementation of consistent international standards seems important, especially for the governance of high-risk technologies that can affect social and environmental systems at a global level.

This raises various issues related to social justice. For instance, what are the consequences of the use and implementation of 'international' regulatory frameworks for low- to middle-income countries? Which forms of resistance and alternative standardization are emerging in response to the circulation of technology standards from high-income countries? Is there a way to combine the costly and technology-intensive processes of international governance for gene drive research with the realization of local economic and innovation opportunities in less wealthy countries?

In the literature on the global governance of the life and health sciences there have been widespread calls for an increased awareness of global inequalities, national particularities, as well as differences in institutional, scientific, and regulatory cultures (Wahlberg et al., 2013; Dove and Özdemir, 2014; Sleeboom-Faulkner, 2016; Liu, 2017; Hauskeller et al., 2017). These studies unanimously reject the notion of a 'one-size-fits-all' approach to global ethical governance by pointing to the need to acknowledge variability in ethical prioritization and the significance of local jurisdictions and socio-cultural contexts (Liu, 2017), the 'Western' origins of many bioethical concepts and procedures (Wahlberg et al., 2013), the role of global scientific governance in fostering new forms of empire and imperial rule (Jasanoff, 2014; Harding, 2015), and the impact of large corporations in shaping international regulatory norms (Abraham, 2002).

A concern with social justice is a common denominator in many of these studies. Social justice, as Mamo and Fishman have pointed out, 'is a public matter focussed on common human interests, equitable distribution of social goods, resources, and opportunities, and a commitment to fostering empowered ... participation' in political and other processes of social and economic life (Mamo and Fishman, 2013). A central aspect of the social justice concept in relation to global technology governance is that the use of international regulatory rules is closely linked to processes of stratification and exclusion (Thévenot, 2009). As stated by Timmermans and Epstein (2010), the adoption of international standards introduces far-reaching changes to local structures and creates new forms of inclusion and hierarchies. Locally evolved research, innovation, and corporate practices are often delegitimized and devalued, and in some cases completely shut down (Timmermans and Epstein, 2010; Birch, 2012).

Considering these factors, calls for the co-development and local adaptation (or indigenization) of 'global' standards and governance models in relation to locally specific conditions, divergent cultural and socio-economic realities, as well as contrasting ethical priorities seem important and well justified. From such a perspective, the global governance of new (bio) technologies does not imply the imposition of unified 'global' standards, but rather refers to a partial convergence of regulatory approaches based on the independent development of national regulations and guidelines (cf. Arcidiacono et al., 2012).

The practical challenge of implementing such an approach aside, one key question remains: In which ways are calls for flexible and context-sensitive international governance frameworks compatible with high-risk biotechnologies that have global and species-level stakes, such as gene drive technology? After all, if such technologies spiral out of control, their social and environmental effects are likely to affect vast areas of the world, far beyond national regulatory borders.

This is the central question that I address in this chapter. I will show that in light of the differing types of risks of gene drive and stem cell research, very different dilemmas and challenges for the global governance of these two technology fields arise. This situation raises important questions with regard to the theorization, role, and limitations of the social justice concept in the context of technology governance – especially since the realization of social justice for some can result in increased risk and new forms of injustice for others.

The chapter starts with a theoretical reflection on the governance of new technologies, with a particular focus on the management of technology risks and the realization of social justice. In the empirical section I will examine the dilemmas and tensions that exist between these two dimensions, by focussing on the governance of stem cell research on the one hand, and emerging governance models for gene drive research on the other hand. In the conclusion I will discuss the main findings and reflect on the limitations of the social justice concept in analysing the challenges and conflicts that underlie processes of technology governance.

Global technology governance: a conflict between risk management and social justice?

The development of regulatory frameworks to govern the evaluation, market approval, and routine use of new technologies unfolds in relation to a set of conflicting objectives and priorities. These are, among others: (i) the economic, scientific, and social potential of new technologies (Vermeulen et al., 2012), (ii) whether a new technology corresponds to public interests and is acceptable to citizens (Benz and Papadopoulus, 2006), (iii) the weighing of technology risks against benefits (Van Zwanenberg et al., 2011), and (iv) issues of social justice (Reardon et al., 2015). Regulatory approaches for evolving technologies can have both an enabling and a disabling effect on technology innovation (Sleeboom-Faulkner et al., 2016). Governance frameworks, depending on the rules and procedures they employ, can influence levels of experimental freedom and the methodologies and criteria through which the safety, efficacy, and utility of a new technology are established (Rosemann and Chaisinthop, 2016). These aspects have a huge impact on the costs of the research and development process, and the market price of new technology applications. In doing so, they shape the ways and the extent to which citizens and technology users are protected from potential harms and risks (Sleeboom-Faulkner, 2016). By influencing the market price of new technologies and the conditions under which novel technology forms can be used, governance mechanisms also have an impact on who will be able to gain access to these technologies and their potential benefits (Mamo and Fishman, 2013). In this chapter I am concerned in particular with the following two criteria: the management of technology risks and the realization (and/or prevention) of social justice.

Risk management versus social justice

Questions about benefits and risks – as the US National Academy of the Sciences' report 'Gene Drive on the Horizon: Advancing Science, Navigating Uncertainty and Aligning Research with Public Values' has pointed out – address primarily the ontological, the material, and the societal impact of new technologies (National Academies, 2016). Questions regarding scientific benefits, for example, examine what kinds of breakthrough can be achieved; which kinds of solutions, products, and profits can be realized; and which types of industries, markets, and employment possibilities are likely to emerge. Questions concerning risks, on the other hand, probe into the possible impact of new technologies on the health and safety of citizens and other species, environmental sustainability, and the potential a new technology has for malignant applications and existential risks (Bostrom, 2013; National Academies, 2016). This also includes questions regarding so-called 'rogue' applications, which are likely to emerge in regulatory grey areas and which may increase risks or enable potentially harmful applications of a new technology (Van Zwanenberg et al., 2011). Questions about social justice, in contrast, are more concerned with the social impact of new technologies, in terms of unequal access and marginalization (Reardon et al., 2015). Social justice, as Reardon has pointed out, 'offers a space to think about others' (Reardon, 2013, 4). The concept allows for the examination of processes of inclusion and exclusion and the underlying mechanisms that create these dividing lines – in relation to variables such as gender, age, ethnicity, education, socio-economic status, cultural and religious values, geographical regions, cross-regional inequalities, and others.

Issues related to social justice in science and technology research concern questions such as who will get access to new technologies, and which social groups or geographic regions will be more exposed to risks and harms than others (National Academies, 2016). Issues of social justice also involve enquiries as to who makes the decisions about these technologies, and who will be protected from risks and harms and who not (Van Schomberg, 2013; Stilgoe et al., 2013). Furthermore, questions of social justice can also focus on the situation of knowledge producers such as scientists, corporations, and universities. In this case relevant questions pertain to who is able to conduct research into a new technology field, and who possesses the resources, expertise, and skills to translate this research into new products and economic profits. And relatedly, who is excluded from these processes and for what reasons (Birch, 2012; Sleeboom-Faulkner et al., 2016). A closely related set of questions is, as already indicated above, *which kind of factors shape such processes of inclusion and exclusion and what role global asymmetries and differentials play in this regard?* (Van Zwanenberg et al., 2011; Sleeboom-Faulkner, 2016). In the following sections of this chapter I will illustrate that a range of fundamental dilemmas exists between effective forms of technology governance and the realization of social justice. I will do so first by referring to previous work on stem cell research and second by commenting on the case of gene drive technology. In the following sections it will become clear that by looking at the specific types of risks of these two technology fields, very different types of questions and tensions emerge, which have profound implications for processes of global governance.

Stem cell research: an emerging medical and regulatory paradigm

The clinical testing and market approval of stem cell-based treatments in the European Union and the USA occur mainly through a multiphase clinical trial system, with the randomized controlled trial (RCT) as gold standard. Drug regulatory authorities in these countries demand compliance with international best practice and evidence-based medicine (EBM) standards. The commitment to the methodological safeguards of the EBM system is also reflected in the 2015 guidelines of the International Society for Stem Cell Research (ISSCR), which is the global scientific elite organization for stem cell research. The 2016 ISSCR guidelines demand a 'cells-as-drugs' approach for stem cell treatment that involves rigorous regulatory oversight. It calls for 'many years of preclinical research and clinical testing prior to the maturation of experimental treatments into an accepted standard of medical practice' (International Society for Stem Cell Research, 2008). It demands, furthermore, that stem cell-based innovations outside of the formal multiphase trial process are justified only in exceptional circumstances in some very limited cases (International Society for Stem Cell Research, 2016).

In practice, however, a more diversified situation has emerged. The striving for international harmonization – exemplified by the efforts of the ISSCR, the Advanced Therapy and Medicinal Products (ATMP) Cluster of the US Food and Drug Administration (FDA), the European Medicines Agency, and Health Canada – is challenged by an increasing trend of regulatory diversification. Three central dynamics of diversification can be observed.

A first dynamic is the emergence of a growing number of regulatory exceptions and exemptions in the USA and the European Union (Faulkner, 2016). These exceptions/exemptions facilitate processes of clinical translation, and sometimes commercial clinical applications (Salter et al., 2015), outside of the multiphase trial system but within the confines and review procedures of national regulatory agencies (Faulkner, 2016). More recently, in the USA the passing of right-to-try laws in more than thirty states has permitted the administering of unproven, experimental medical interventions to patients outside of the review of the US FDA. The passing of the

21st Century Cures Act in the USA in December 2016 has introduced additional possibilities to avoid going through large-scale phase III trials, which is expected to accelerate market approval of new medicines, including stem cell and regenerative medicine products (Hogle and Das, 2017). Salter et al. (2015) have interpreted the emergence of these regulatory exceptions as a process of 'hegemonic adaptation' through which nation states try to correspond to growing demands for accelerated and more widespread access to investigative medicines, 'whilst still retaining [the] essential character and power relationships' of a state-led model of drug development (2015, 158).

A second dynamics of diversification is the flexible enforcement of regulatory standards in some countries, which has enabled the continuous provision of experimental for-profit interventions with stem cells outside the regulatory apparatus of the state. In China and India, for example, unproven for-profit stem cell interventions have been tolerated by state authorities for many years despite the fact that these interventions are now in both countries officially prohibited (McMahon, 2014; Rosemann et al., 2016). Professional societies in these countries have promoted alternative standards and research methodologies that encourage physician-based forms of clinical experimentation with stem cells, independent of the review of regulatory agencies (Rosemann and Chaisinthop, 2016). It is an open question whether the flexible enforcement of regulatory standards in China and India represents a deliberate political strategy that will enable domestic research and innovation practices (Sleeboom-Faulkner et al., 2016), or whether it reflects the lack of regulatory resources and infrastructures or simply a lack of interest, in the light of other more challenging healthcare priorities (Tiwari and Raman, 2014).

A third process of regulatory diversification is the complete abandoning of the EBM and multiphase trial system. This has recently happened in Japan. Following a regulatory reform in 2014, the Japanese system allows now for conditional market approval of stem cell interventions after a small-scale phase I trial with only ten patients or even fewer, provided the applied treatments appear effective and safe (Sipp, 2015; Lysaght, 2017). Once the Japanese Pharmaceuticals and Medical Devices Agency has provided conditional approval for a stem cell intervention, clinical researchers and corporations have the possibility of seeking conditional approval for up to seven years (Sipp, 2015). It is not unlikely that other countries will follow the Japanese regulatory model, or at least create additional types of regulatory exceptions in which (conditional) market approval of stem cell therapies can be granted without preceding phase I–III trials.

Dilemmas between risk management and social justice – the stem cell case

It is of particular interest with regard to this chapter that both the opposition to and the support for the EBM and the multiphase trial system – which is still at the heart of the US-European regulatory approach for stem cell medicine – are based on varying and often contradictory forms and claims of social justice. I will now illustrate this from the perspectives of regulatory authorities, scientists, and medical entrepreneurs, and, third, patients.

Social justice from the perspective of regulatory agencies

Throughout human history decisions on how to treat sick or disabled persons were overwhelmingly based on personal experience, and on the observation of individuals or small groups of patients. With the advance of modern biomedicine, and in the wake of severe medical accidents and dramatic cases of exploitative medical experiments, clinical research took an increasingly controlled and systemic form. Drug regulatory authorities became the central custodians of this approach, whose scientific and ethical principles were institutionalized in law, mandatory best practice standards, and enforced by government agencies (Roman, 2014). A central aim of the multiphase trial and EBM system has been to assure that patients can access safe and effective medicines. A key role of drug regulatory authorities such as the US FDA is to prevent the premature marketization of new medical interventions which may pose health risks to patients, and expose them to financial exploitation and therapies that do not work. Hence, from the perspective of regulatory agencies, the protection of patients from irresponsible experimentation as well as scrupulous companies and scientists can be seen as an attempt to realize one particular type of social justice: the right of patients to efficient and safe medicines and healthcare.

Social justice from the perspective of scientists

For many scientists and small- to middle-sized corporations the rules and protective role of drug regulatory authorities pose a practical challenge. The need to comply with international standards, EBM principles, and the multiphase trial system exceeds the financial and infrastructural resources of these stakeholders – which often has an inhibiting and marginalizing effect. While this is a frequent complaint in high-income countries, this problem is magnified in low- to middle-income countries. International technology regulation, from this perspective, can form a crucial mechanism of exclusion which prevents the participation of stakeholders in clinical innovation processes, for whom adherence to international standards and the multiphase trial system is too expensive and beyond reach. In the stem cell field this has resulted in widespread forms of resistance and the development of alternative and less stringent methodological and ethical standards that legitimize for-profit interventions with stem cells in an early experimental stage, where reliable data for safety and efficacy do not yet exist. As documented elsewhere (Rosemann and Chaisinthop, 2016), for many researchers and medical entrepreneurs in low- to middle-income countries, opposition to the international regulatory system is an attempt to reverse what they perceive as a form of global injustice (Rosemann and Chaisinthop, 2016). From the viewpoint of these stakeholders, attempts to criticize, deconstruct, and reshape the international EBM system as a shared global standard are a critical means to reclaim their position as participants in the clinical innovation process. Resistance to the authority and global reach of drug regulatory agencies in high-income countries is a central instrument in re-establishing their 'right to innovate and make profits' without being pushed aside by regulatory norms developed by elite scientists in high-income countries (Sleeboom-Faulkner, 2016). The restoration of social justice, from the perspective of these actors, is to break the global hegemony of EBM standards and international rules which delegitimizes their activities and which brands their work as 'rogue', 'irresponsible', or 'unscientific' (Salter et al., 2015; Sleeboom-Faulkner, 2016).

Social justice from the perspective of patients

Patients who seek access to stem cell treatments are confronted with the following dilemma. On the one hand, patients want access to safe and effective medicines, which is in line with the aim of drug regulatory authorities. On the other hand, if treatments are still in the development pipeline, patients often want rapid access (Lindvall and Hyun, 2009). The mere possibility that a candidate treatment will help is a pull factor for many prospective patients even if this means accepting greater risks and the fact that the treatment may be ineffective. However, the widespread wish among patients to obtain rapid access to experimental interventions clashes with the slow process of the multiphase trial system. As a result, patients are frequently prevented from accessing these experimental interventions. Many patients and patient organizations consider the lack of access to these possibly helpful interventions as a form of injustice, in which individual decisions are restricted through a 'patronizing' apparatus of experts and regulatory institutions (Madden and Conko, 2013). In recent years, patient activists and political lobby groups have increasingly sought to change this situation (Bianco and Sipp, 2014). By advocating for their 'right to try' and the 'freedom to choose', these groups seek to realize what they perceive as a fundamental form of social justice (ibid.), even if this means exposing their bodies to risks that regulatory agencies consider unacceptable.

As these examples show, varying claims of 'social justice' are acting upon each other in the stem cell field, in a complex and contradictory relationship. For regulatory agencies, the realization of social justice lies embedded in the protection of patients from accessing treatments that are inefficient or unsafe. For many scientists and corporations in low- to middle-income countries, on the other hand, demands for social justice are implicated in the struggle to innovate and make profits, without the restricting rules of professional and regulatory institutions from high-income countries. Large numbers of patients and patient organizations, in turn, seek to realize social justice by advocating for increased autonomy in trying out non-approved candidate treatments, even if this means to be exposed to potentially dangerous and ineffective medical interventions. Of interest in this regard is the fact that the model of risk management (and the demands for social justice that are embedded in this model), which is handled by regulatory agencies in high-income countries, conflicts with both the ideas of many scientists and medical entrepreneurs, as well as the demands of a large number of patients. Both of these latter groups deliberately accept higher risks and argue for a less stringent model of risk management in order to realize what they consider to be a greater level of social justice: unrestrained possibilities for localized forms of medical innovation and a greater level of freedom to choose and try out unproved but potentially life-saving candidate treatments. In the gene drive field, the dilemmas and tensions between risk management and social justice differ in important respects from stem cell research. However, here too conflicts between standards and regulatory controls developed by professional elite groups in high-income countries, and the interests of less well-endowed researchers and corporations, are playing an important role.

Gene drive research: shaping the environment, sculpting evolution?

The term 'gene drive' refers to a form of genetic engineering that allows for the distribution of genetic modifications through biological populations at a pace that is significantly faster than natural reproduction and older forms of genetic engineering. Once a genetic change has been introduced into an organism, this modification then 'drives' through a biological population until each organism has inherited the change. This means that the release – accidental or intentional – of a single or a few modified organisms could quite possibly genetically change all populations of a species through the world (Esvelt, 2016). Gene drive technology is associated with important benefits such as agricultural applications (for example the control of invasive species), environmental applications (the conservation of species that would otherwise become extinct), as well as healthcare applications (especially the control of disease infections such as malaria, dengue fever, and Zika) (Saey, 2015).

However, there are also numerous risks related to gene drive technology. These are, first, irreversible alterations to the ecosystem, with the potential to interrupt the biological balance between species and to prevent environmental sustainability (Oye et al., 2014). Second, unintended consequences could occur such as the possibility of gene drives moving into other species (ibid.). Third, accidental changes of ecosystems and the natural environment can be expected to affect the functioning of human societies, and to affect human well-being (ibid.). Fourth, there is the possibility of the misuse (or dual use) of gene drive applications, for instance in the context of military applications of gene drive technology, or in forms of bioterrorism. Indeed, there have been various reports that the Pentagon, and the military departments of other nations, are now investing in gene drive research (Begley, 2015).

Governance of gene drive research and applications

The risks of gene drive technology do not stop at national borders. Harmful applications can affect various world regions, and potentially the entire planet. For this reason there is a widespread consensus that international governance frameworks are urgently needed to better control this emerging technology field, and to enable responsible forms of technology use. At present, various multi-lateral fora have published recommendations for such international standards (Oye et al., 2014; Akbari et al., 2015; National Academies, 2016). Most participants in these fora and the involved scientific groups come from high-income countries. Scientists and regulators from low- to middle-income countries – including scientifically advanced, middle-income countries in which gene drive research can be expected to flourish and to be conducted in field experiments – have been under-represented.

While actual regulatory instruments and solutions for gene drive applications are only gradually emerging, it is becoming clear that emerging governance forms will be expensive, technology-intensive, and will require extensive and well-functioning regulatory and scientific infrastructures. Each phase of the research process (basic research in the laboratory, controlled field trials, and later environmental release for gene drive species) requires a broad range of security measures and extensive, longitudinal forms of risk assessment (Oye et al., 2014). The implementation of appropriate safeguards, risk evaluation, and long-term monitoring of the potential environmental impact of released gene drive organisms involves complex inter-disciplinary collaborations and forms of mathematical modelling that require time and significant resources (Akbari et al., 2015; Esvelt, 2016). The development of transnational projects – such as the malaria gene drive project headed by Austin Burt at Imperial College London, which targets the release of gene drive modified mosquitoes to eradicate malaria in sub-Saharan Africa – necessitates the building of a complex international infrastructure and highly structured long-term planning. This involves the development of multi-national research teams, the development of extensive training and capacity building in African countries (in which environmental release will one day take place), as well as substantial investments in local scientific infrastructures (Akbari et al., 2015). It also includes the conduct of long-term public engagement with multiple local communities in all involved countries, including neighbouring countries that might be affected by the release of gene drive organisms indirectly (ibid.).

These measures require intensive funding for periods of more than one decade, as well as highly trained and committed organizational staff. But expenses and work also continue after the environmental release of gene drive organisms. Security measures that have been requested by international committees are the long-term genetic and environmental monitoring of other species in ecosystems (that might be affected by the release of genetically modified [GM] organisms), and the engineering of so-called reverse gene drives (National Academies, 2016). A reverse drive, provided it works, would have the potential to overturn the effects of released gene drive organisms, if these produced negative or undesired outcomes (Oye et al., 2014).

Challenges to the governance of gene drive research in low- to middle-income countries

For researchers, public funding bodies, and corporations in low- to middle-income countries, it will be difficult to adopt and implement these safeguards. Based on my research into the governance challenges of stem cell research in low- to middle-income countries (Rosemann, 2014; Rosemann and Chaisinthop, 2016; Rosemann et al., 2016), it is quite reasonable to expect that the adoption and implementation of these emerging, expensive, and complex regulatory solutions for gene drive technology will be difficult to realize – and, in fact, is unlikely to happen in many countries, at least at a truly consistent level.

Based on insights from the stem cell field, four developments can be expected. First, due to a lower level of available financial, technological, and infrastructural resources in low- to middle-income countries, it is probable that resistance to costly international standards will emerge. This is likely to result in the development of less costly regulatory alternatives that correspond to local circumstances and priorities. Second, the flexible or lenient implementation of regulatory rules can be expected. A central insight from the governance of clinical stem cell research was that scientifically advanced middle-income countries such as China and India have adopted regulatory frameworks that – at least on paper – have in many regards embraced regulatory standards from high-income countries. In practice, however, the implementation of regulatory rules has proved to be lenient and flexible. While official standards have been used for domestic elite science projects and international collaborations, outside of these projects regulatory standards have often not, or only partly, been implemented (Rosemann et al., 2016). Similar developments can be expected in the gene drive field. The selective, flexible, or lenient implementation of regulatory standards is in many countries exacerbated by the lack of consistent regulatory and scientific infrastructures. This is particularly the case in large, population-rich countries with a high number of geographically dispersed scientific institutions and corporations. The consistent and long-term monitoring of research practices across these large numbers of institutions is challenging, and offers opportunities for more informal forms of research and technology application practices outside the public eye and away from the control of regulatory agencies. A closely related point is that divergent economic and political priorities as well as differing availability of resources are likely to result in the emergence of informal commercial applications and regulatory grey areas. Evidence of this comes not only from the stem cell field, but in particular also from social science studies on the agricultural use of GM crops. As Van Zwanenberg et al. (2011) have pointed out, in both Latin American and Asian countries large-scale informal markets with GM crops have emerged that in many regards have conflicted with formal regulatory rules but which formed a more affordable alternative for local farmers. This is a third development that can be expected in the gene drive field. A fourth point relates to the existence of different regulatory frameworks for military and public research institutions. Considering the military potential of gene drive applications (for example, suppressing pollinators or to give innocuous insects the ability to carry dangerous diseases [Begley, 2015]), military research institutions in various countries can be expected to launch gene drive research programs which at a later stage may also involve field experiments outside the reach of public scrutiny. In China, for instance, in the stem cell field, different regulatory requirements in military institutions have resulted in the continued provision of experimental stem cell treatments, while in public research institutions these interventions were prohibited.

Together, these factors are likely to increase shared global risks of gene drive applications, and to constitute a significant challenge to the consistent control of this technology field. While the above-mentioned scenarios are at present merely hypothetical, some of these developments can easily become reality in the near future. For policy-makers and scientists, it will be vital to take these possibilities into account, to contemplate the risks of these possible developments, and to integrate these insights into international debates and policies. One of the key challenges for the consistent governance of gene drive (and other gene editing) technology is that with the arrival of the CRISPR-Cas system, genetic modifications of living organisms are relatively easy to perform and at a low cost. As Jennifer Doudna, one of the inventors of the CRISPR technology, has remarked, in principle a graduate student with access to basic laboratory equipment could engineer a GM organism and release it into the environment (Baltimore et al., 2015). This could theoretically induce gene drives in biological populations, entirely without external controls, approval, or monitoring.

Dilemmas between risk management and social justice – gene drive research

In the domain of clinical stem cell research, but also in the gene drive field, resistance to – and the reformulation of – international regulatory norms is in many regards understandable. This is true, in particular, from the perspective of researchers and start-up companies that do not possess the financial, technological, or infrastructural means to comply with international standards. In order to prevent marginalization and to open up niche areas of innovation and profit-making, these stakeholders oppose inter-national standards and create their own rules. As Sleeboom-Faulkner has shown in a case study of a stem cell company in China, participation in the recognition and reward systems of the international arena of 'high profile science' (e.g. publications in high-ranked academic journals, prestigious research grants, academic prizes and awards, etc.) is for many researchers unimportant or secondary as long as there are possibilities to reap rewards, gain recognition, and reach entrepreneurial success at a national or regional level (Sleeboom-Faulkner, 2016). This is the case not only for researchers in low- to middle-income countries but also for many scientific and medi-cal entrepreneurs in high-income countries. As suggested in a recent article with Nattaka Chaisinthop, conflicts surrounding international standards are understudied aspects of wider and more complex dynamics of global competition around new economic opportunities, employment possibilities, and scientific breakthroughs (Rosemann and Chaisinthop, 2016). Opposi-tion to international standards and regulatory frameworks, from this per-spective, is a strategic attempt to revalue, legitimize, and garner recognition for forms of research and market use that would otherwise be precluded and considered unacceptable by many 'elite' scientists. As mentioned above, in the clinical stem cell field, but also in other areas of medical research, this struggle for recognition is frequently cast in a terminology of 'rights' and 'social justice'. The rejection and re-articulation of international research standards is framed as a crucial means to realize local innovation practices and to reassert the right for local innovation and profit-making, without being pushed away by the regulatory norms from the scientific elites in high-income countries (Sleeboom-Faulkner 2016).

The divergent risks of stem cell and gene drive research: implications for governance

I will now briefly reflect on the implications of these forms of resistance to in-ternational standard regimens. While opposition and alter-standardization may have been particularly pronounced in the clinical stem cell field, similar dynamics have been observed in GM agriculture (Van Zwanenberg et al., 2011). Considering the widespread global interest in gene drive research, its broad range of possible applications, and for-profit potential, there is good reason to expect that comparable developments (i.e. demands for regulatory variation and the development of regulatory frameworks that are sensitive to local circumstances and differences in available resources) will emerge in the gene drive field. Considering that international standards create new boundaries of inclusion and exclusion among individuals, groups, practices as well as opportunities (Thévenot, 2009), and the fact that researchers and corporations in low- to middle-income countries are often dispropor-tionately affected by the implementation of 'hegemonic' global technology standards, the circumvention and the wish to renegotiate international norms are in many regards understandable. But how acceptable is regulatory variation and the adoption of more lenient and flexible technology standards, in the light of the specific technology risks of stem cell and gene drive research?

If we look more closely at the types of risk that these two technology fields create, it becomes clear that very different dilemmas exist between risk management and 'social justice'. These differences have profound implications for the global governance of these technology areas. The risks of clinical stem cell research play out primarily at the level of individual patients and their families. Possible risks include medical malpractice, adverse effects, or the provision of ineffective treatments. Correlated risks are negative mental health impact and the risk of financial exploitation, both of which may affect the well-being of the wider family or friends. Most people who undergo experimental stem cell interventions are typically consenting individuals (albeit the reason for their 'consent' is likely to be desperation), who are in most cases aware of the risks to which they are exposed. While it is clear that the risks of stem cell treatments require careful management at the level of human societies as a whole, the risks of experimental stem cell interventions are confined to a relatively 'small' segment of the human population. I by no means intend to trivialize or downplay the risks of clinical stem cell interventions here. However, it seems safe to argue that the magnitude of risks in the gene drive field is of a larger magnitude, and also that risks reach far beyond those of humans. In the gene drive field, risks are not restricted to (consenting) individuals and specific groups of patients but are of a much more collective nature. Risks in the gene drive field have the potential to affect both ecosystems and the natural environment, as well as humans and human societies. The most extreme risks of gene drive research, such as bioterrorism or military abuse, bear the potential to affect and kill large numbers of humans, potentially in multiple world regions (Caplan et al., 2015). There is also the risk of the extinction of other biological species as an unintended side effect of gene drive applications in one particular species. An additional problem is that – due to the large number of factors that interact in global ecosystems – the actual risks of gene drive applications are much more difficult to define, predict, and measure. The impact of field trials and the environmental release of gene drive modified species may be unexpected and indirect, and unfold over extended periods of time (Esvelt, 2016). In fact, reliable forms of ecological and societal risk assessment for this technology field are only emerging and there is disunity among experts on the kinds of evidence that are needed to support the release of gene drive applications into the environment (Oye et al., 2014; Akbari et al., 2015; Esvelt, 2016).

These different risk profiles have fundamental implications for the governance of gene drive applications. In the clinical stem cell field, a higher level of regulatory flexibility and variation – that adjusts to local circumstances and unequal availability of resources – makes opposition to, circumvention and re-articulation of international standards more acceptable than in the gene drive field. While the risks related to clinical stem cell interventions require careful management, one can empathize with the wish of researchers and start-up corporations to break with the rigorous methodology of the multiphase trial and EBM system and to develop regulatory alternatives that are more affordable. In the gene drive field a different situation exists. The more collective, global, and often-indirect risks of gene drive technology pose a significant challenge to the acceptability of claims for 'social justice', 'inclusion', or 'the right to innovate' by researchers who may be excluded by international regulatory standards. Hence, in the gene drive field demands for 'regulatory variation' or the 'flexible handling' of international regulatory mechanisms seems more problematic and extremely difficult to justify. Considering the fact that the biological and social effects and risks of gene drive applications do not stop at national borders, regulatory variability across countries poses a significant challenge. If institutions, regions, or countries opt for more lenient and flexible regulations, potentially negative effects are not restricted to local contexts, and are likely to pose a danger to biological species and human societies in larger parts of the world, potentially at a global level.

Conclusions

In this chapter I have explored dilemmas and tensions related to the management of risks and the realization of social justice in the context of the international governance of stem cell and gene drive research. Considering recent processes of regulatory diversification, resistance to international technology standards, and the development of regulatory alternatives in the stem cell field, I have examined whether similar developments can also be expected in the gene drive field, and what the consequences of these developments could be. A rejection of a 'one-size-fits-all' approach to global technology governance is now a widely shared position in the sociological and anthropological literature on global technology governance. From an egalitarian, inclusion-oriented, social justice perspective the rejection of globally unified governance forms (that inevitably include forms of domination, bias, and often disruptive interventions in local contexts) is of course well justified. A more open, flexible governance approach that leaves space for regional variation and indigenization makes a lot of sense. However, as I have set out to ask in the Introduction, in which ways are calls for flexible and context-sensitive international governance frameworks compatible with high-risk biotechnologies that have global and species-level stakes, such as gene drive technology?

In an attempt to answer this question, I have argued that – due to the collective, global, and often-indirect risks of gene drive technology that will affect not only humans and human societies but also the natural environment and ecosystems – the acceptability of claims for social justice, inclusion, and more lenient regulations is much more difficult to justify and accept than in the stem cell field. The shared global risks of gene drive research, it seems to me, necessitate the consistent implementation of internationally 'harmonized' regulatory protocols and standards, and will in many respects leave little space for local variation.

The adoption of more flexible, variable, and less stringent regulatory standards that can currently be observed in the stem cell field (and which are indicative of a broader trend of deregulation in medicine research) is in any case difficult to justify in the gene drive field. In practice, as gene drive technology develops and moves towards environmental release, divergent regulatory safeguards between countries will probably be unavoidable. But deviation from consistent and stringent regulatory controls of this technology field will increase global risks and is likely to give rise to developments and problems that at present are hard to foresee. This gives rise to a dilemma: while it will be important to explore what kinds of exclusion are likely to emerge in the context of emerging international governance frameworks for gene drive technology, it will be equally important to understand what costs could come with flexible regulation, or with the deviation from recognized international standards – at a local, regional, and global level.

This points to clear limitations of a social justice perspective as a generating principle in the shaping of technology governance. While the concept of social justice – and its promise to empower and to identify and prevent processes of exclusion – is of vital importance, there are also related drawbacks. As mentioned by Reardon, a simultaneous strength and danger of the justice concept is its potential to gather people together, to mobilize, and to initiate collective and political action (Reardon, 2013). As this chapter has shown, in the stem cell field calls for justice have been mobilized for varying (and sometimes conflicting) purposes – by patients, researchers, regulatory institutions, and corporations. Demands for less stringent standards, more early and widespread access to unproven clinical interventions, and resistance to the multiphase trial system have resulted in the lowering of methodological standards and in the widespread provision of clinical applications whose effectiveness and safety can often be questioned. These developments may have lowered drug development costs and created new economic opportunities, and they have possibly helped some patients, but they have also exposed patients to greater health risks and the risk of financial exploitation. Accordingly, demands for social justice that underlie current processes of deregulation in the stem cell field (and other areas of medicine) can easily result in new forms of injustice. The line between the realization of justice and the institutionalization of new forms of injustice is often thin, and closely intertwined. In the gene drive field in any case, regulatory variability and deregulation in the name of justice seems hard to justify.

Be this as it may, consistent implementation of crucial regulatory safeguards at a global level can hardly be expected, especially if a larger number of researchers and corporations will get involved in gene drive research. In the end, some degree of regulation and control will still be better than none, whatever the implications of gene drive applications for the global biosphere may be.

Acknowledgements

This article has benefited from research support provided by the ERC (283219), the ESRC (ES/I018107/1), and the Wellcome Trust (204799/Z/16/Z). I would like to thank Anja Pichl, Arne Manzeschke, and Christine Hauskeller for their valuable feedback.

References

Abraham, J. (2002). The pharmaceutical industry as a political player. *The Lancet*, 360(9344), pp. 1498–1502.

Akbari, O. S., Bellen, H. J., Bier, E., Bullock, S. L., Burt, A., Church, G. M., Cook, K. R., Duchek, P.Edwards, O. R., Esvelt, K. M. and Gantz, V. M. (2015). Safeguarding gene drive experiments in the laboratory. *Science*, 349(6251), pp. 927–929.

Arcidiacono, J. A., Blair, J. W., and Benton, K. A. (2012). US Food and Drug Administration International Collaborations for Cellular Therapy Product Regulation. *Stem Cell Research & Therapy*, 3(5), pp. 38–43.

Baltimore, D., Berg, P., Botchan, M., Carroll, D., Charo, R. A., Church, G., Corn, J. E., Daley, G. Q., Doudna, J. A., Fenner, M. and Greely, H. T. (2015). A prudent path forward for genomic engineering and germline gene modification. *Science*, 348(6230), pp. 36–38.

Begley, S. (2015). Why the FBI and Pentagon are Afraid of this New Genetic Technology. *Statnews*, [online] (12 Nov.). Available at: www.statnews.com/2015/11/12/gene-drive-bioterror-risk/ [Accessed 22 Sept. 2017].

Benz, A. and Papadopoulos, Y. (2006). Actors, institutions and democratic governance: comparing across levels. In: A. Benz and Y. Papadopoulos, eds., *Governance and Democracy. Comparing National, European and International Experiences*, New York: Routledge, pp. 273–295.

Bianco, P. and Sipp, D. (2014). Sell help not hope. *Nature*, 510(7505), pp. 336–337.

Birch, K. (2012). Knowledge, place, and power: geographies of value in the bioeconomy. *New Genetics and Society*, 31(2), pp. 183–201.

Bostrom, N. (2013). Existential risk prevention as global priority. *Global Policy*, 4(1), pp. 15–31.

Caplan, A. L., Parent, B., Shen, M. and Plunkett, C. (2015). No time to waste – the ethical challenges created by CRISPR. *EMBO Reports*, 16(11), pp. 1421–1426.

Dove, E. S. and Özdemir, V. (2014). Glocal bioethics: when international IRB collaboration confronts local politics. *The American Journal of Bioethics*, 14(5), pp. 20–23.

Esvelt, K. (2016). An analysis of gene drive risks and safeguards, [online]. Available at: www.sculptingevolution.org/genedrives/safeguards [Accessed 22 Sept. 2017].

Faulkner, A. (2016). Opening the gateways to market and adoption of regenerative medicine? The UK case in context. *Regenerative Medicine*, 11(3), pp. 321–330.

Harding, S. (2015). Postcolonial and feminist philosophies of science and technology. In: P.K. Nayar, ed., *Postcolonial Studies: An Anthology*, Hoboken (New Jersey) and Oxford: Wiley-Blackwell, pp. 533–552.

Hauskeller, C., Baur, N. and Harrington, J. (2017). Standards, harmonization and cultural differences: examining the implementation of a European stem cell clinical trial. *Science as Culture*, pp. 1–26, doi:10.1080/09505431.2017.1347613

Hogle, L. and Das, A. (2017). The social production of evidence: regenerative medicine and the U.S. 21st century cures act. *Regenerative Medicine*, 12(6), pp. 8–12.

International Society for Stem Cell Research. (2008). ISSCR Guidelines for the Clinical Translation of Stem Cells, [online]. Available at: www.isscr.org/docs/default-source/all-isscr-guidelines/clin-trans-guidelines/isscrglclinicaltrans.pdf?sfvrsn=6 [Accessed 8 June 2019].

International Society for Stem Cell Research. (2016). Guidelines for Stem Cell Research and Clinical Translation, [online]. Available at: www.isscr.org/docs/default-source/all-isscr-guidelines/guidelines-2016/isscr-guidelines-for-stem-cell-research-and-clinical-translationd67119731dff6ddbb37cff0000940c19.pdf?sfvrsn=4 [Accessed 8 June 2019].

Jasanoff, S. (2014). Biotechnology and empire: the global power of seeds and science. In: M. Mayer, M. Carpes and R. Knoblich, eds., *The Global Politics of Science and Technology–Volume 1*, Berlin, Heidelberg: Springer, pp. 201–225.

Lindvall, O. and Hyun, I. (2009). Medical innovation versus stem cell tourism. *Science*, 324(5935), pp. 1664–1665.Liu, J. A. (2017). Situated stem cell ethics: beyond good and bad. *Regenerative Medicine*, 12(6), pp. 587–591.

Lysaght. T. (2017). Accelerating regenerative medicine: The Japanese experiment in ethics and regulation. *Regenerative Medicine*, 12(6), pp. 32–40.

Madden, B. and Conko, G. (2013). Free to choose medicine. The Federalist Society for Law and Public Policy Studies, [online]. Available at: www.fed-soc.org/publications/detail/free-to-choose-medicine [Accessed 22 Sept. 2017].

Mamo, L. and Fishman, J. R. (2013). Why justice? Introduction to the special issue on entanglements of science, ethics, and justice. *Science, Technology, & Human Values*, 38(2), pp. 159–175.

McMahon, D. S. (2014). The global industry for unproven stem cell interventions and stem cell tourism. *Tissue Engineering and Regenerative Medicine*, 11(1), pp. 1–9.

National Academies. (2016). Gene Drives on the Horizon: Advancing Science, Navigating Uncertainty, and Aligning Research with Public Values, [online]. Available at: http://nas-sites.org/gene-drives/ [Accessed 22 Sept. 2017].

Oye, K. A., Esvelt, K., Appleton, E., Catteruccia, F., Church, G., Kuiken, T., Lightfoot, S. B. Y., McNamara, J., Smidler, A. and Collins, J. P. (2014). Regulating gene drives. *Science*, 345(6197), pp. 626–628.

Reardon, J. (2013). On the emergence of science and justice. *Science, Technology, & Human Values*, 38(2), pp. 176–200.

Reardon, J., Metcalf, J., Kenney, M. and Barad, K. (2015). Science & justice: the trouble and the promise. Catalyst: Feminism, *Theory, Technoscience*, 1(1), pp. 1–49.

Roman, S. (2014). The history of clinical trials. In: T. M. Pawlik and J. A. Sosa, eds., *Success in Academic Surgery: Clinical Trials*. London: Springer, pp. 141–152.

Rosemann, A. (2014). Standardization as situation-specific achievement: regulatory diversity and the production of value in intercontinental collaborations in stem cell medicine. *Social Science & Medicine*, 122, pp. 72–80.

Rosemann, A. and Chaisinthop, N. (2016). The pluralization of the international: resistance and alter-standardization in regenerative stem cell medicine. *Social Studies of Science*, 46(1), pp. 112–139.

Rosemann, A., Bortz, G., Vasen, F. and Sleeboom-Faulkner, M. (2016). Global regulatory developments for clinical stem cell research: diversification and challenges to collaborations. *Regenerative Medicine*, 11(7), pp. 647–657.

Saey, T. (2015). Gene drives spread their wings, *ScienceNews*, [online] (12 Dec.). Available at: https://www.sciencenews.org/article/gene-drives-spread-their-wings, [Accessed 8 June, 2019].

Salter, B., Zhou, Y. and Datta, S. (2015). Hegemony in the marketplace of biomedical innovation: consumer demand and stem cell science. *Social Science & Medicine*, 131, pp. 156–163.

Sipp, D. (2015). Conditional approval: Japan lowers the bar for regenerative medicine products. *Cell Stem Cell*, 16(4), pp. 353–356.

Sleeboom-Faulkner, M. (2016). The large grey area between 'Bona Fide' and 'Rogue' stem cell interventions—ethical acceptability and the need to include local variability. *Technological Forecasting and Social Change*, 109, pp. 76–86.

Sleeboom-Faulkner, M., Chekar, C. K., Faulkner, A., Heitmeyer, C., Marouda, M., Rosemann, A., Chaisinthop, N., Chang, H. C. J., Ely, A., Kato, M. and Patra, P. K. (2016). Comparing national home-keeping and the regulation of translational stem cell applications: an international perspective. *Social Science & Medicine*, 153, pp. 240–249.

Stilgoe, J., Owen, R. and Macnaghten, P. (2013). Developing a framework for responsible innovation. *Research Policy*, 42(9), pp. 1568–1580.

Thévenot, L. (2009). Postscript to the special issue: governing life by standards: a view from engagements. *Social Studies of Science*, 39(5), pp. 793–813.

Timmermans, S., and Epstein, S. (2010). A world of standards but not a standard world: Toward A sociology of standards and standardization. *Annual Review of Sociology, 36*, pp. 69–89.

Tiwari, S. S., and Raman, S. (2014). Governing stem cell therapy in India: regulatory vacuum or jurisdictional ambiguity? *New Genetics and Society, 33*(4), pp. 413–433.

Von Schomberg, R. (2013). A vision of responsible innovation. In: R. Owen, M. Heintz and J. Bessant, eds., *Responsible Innovation: Managing the Responsible Emergence of Science and Innovation in Society*, London: Wiley and Sons, pp. 51–74.

Vermeulen, N., Tamminen, S. and Webster, A., eds. (2016). *Bio-Objects: Life in the 21st Century*. London: Routledge.

Wahlberg, A., Rehmann-Sutter, C., Sleeboom-Faulkner, M., Lu, G., Döring, O., Cong, Y., Laska-Formejster, A., He, J., Chen, H., Gottweis, H. and Rose, N. (2013). From global bioethics to ethical governance of biomedical research collaborations. *Social Science & Medicine*, 98, pp. 293–300.

Van Zwanenberg, P., Ely, A. and Smith, A. (2011). *Regulating Technology*. London: Earthscan.

List of contributors

Antonakaki, Melpomeni
Post/Doc Lab. Engineering Responsibility at Munich Center for Technology in Society, Technical University Munich, Germany
Melpomeni Antonakaki works as scientific associate and doctoral candidate at the Munich Centre for Technology in Society. She has qualified in History of Science and Technology at the National Kapodistrian University of Athens, Greece, but also holds diplomas in Molecular Biology and in History of Philosophy and Psychology from the University of Ioannina, Greece. Her dissertation at the Munich, Center for Technology in Society (MCTS) is situated within the *Engineering Responsibility Lab*, jointly supervised by Prof. Maasen at the Technical University Munich, and Prof. Hilgartner, Cornell University. Inspired by feminist ethno-epistemological approaches to experimental response-ability, Melina focusses on how aspects of 'conduct', intentionality, and bodily performance are being inscribed in software/hardware configurations and how the latter sociotechnical arrangements enact 'proper research subjects' and 'proper biology'.

Beltrame, Lorenzo
Department of Sociology and Social Research, University of Trento, Italy
Lorenzo Beltrame holds a PhD in Sociology. He works at the Department of Sociology and Social Research, University of Trento, where he teaches sociology of innovation and sociology of globalization. Between 2015 and 2017 Lorenzo was Marie Skłodowska-Curie Research Fellow at the University of Exeter, where he worked with Christine Hauskeller on the EU Horizon 2020 funded project Regulating of Umbilical Cord Blood Banking in Europe (REGUCB), studying the role of national and international regulations in shaping the bioeconomies of umbilical cord blood biobanking. His research focusses on the social, institutional, and regulatory aspects of the emergent bioeconomies in the field of stem cell research and technologies.

Fagan, Melinda Bonnie
Department of Philosophy, University of Utah, USA
Melinda Bonnie Fagan is Associate Professor of Philosophy at the University of Utah, where she holds the Sterling M. McMurrin Chair. Her

research focusses on experimental practice in biology (particularly stem cell and developmental biology), explanation, and social epistemology. Before joining Utah's philosophy faculty in 2014, she was Assistant Professor of Philosophy at Rice University (2007–2014) and obtained degrees in History and Philosophy of Science (PhD 2007, Indiana University), Philosophy (MA 2002, University of Texas at Austin), and Biology (BA 1992, Williams College; PhD 1998, Stanford University). Her research in biology focussed on colonial organisms (plants and protochordates) and the evolution of histocompatibility. She is currently working on a view of explanation that looks at concepts of collaboration and interaction.

Ghasemi-Kasman, Maryam
Cellular and Molecular Biology Research Center and Iran Neuroscience Research Center, both at Health Research Institute, Babol University of Medical Sciences, Iran

Maryam Ghasemi-Kasman is Assistant Professor of Medical Physiology at the Babol University of Medical Sciences, Babol, Iran. Her PhD project was about the effect of specific microRNAs as miR-302/367 cluster on direct reprogramming of glial cells into neural progenitors or oligodendrocyte precursor cells. In the next step, the effect of these induced cells on neurogenesis and myelin repair was investigated in animal models. She also examined the reprogramming of glial cells to neurons and the effect of these induced cells on behavioural improvement in the Alzheimer's disease model. Her PhD thesis has provided valuable experiences in the field of cellular reprogramming. These research experiences included molecular biology especially the regulatory mechanisms in microRNAs and stem cell research. She is currently involved in several projects that mainly focus on brain repair in multiple sclerosis and epilepsy animal models.

Hauskeller, Christine
University of Exeter, UK, Department of Sociology, Philosophy and Anthropology

Christine Hauskeller is Professor of Philosophy, teaching moral and social philosophy, bioethics, and Critical Theory. She completed her MA at Goethe University Frankfurt on Main, and her PhD in philosophy at the TU-Darmstadt. Her research is in the philosophy of science and technology, ethics, and social philosophy. Christine conducts empirical philosophical studies on the processes of knowledge production in the life sciences and their intersections with different forms of valuation and normativity. She has published widely on stem cell research for the past twenty years and has led many projects on the subject area funded mostly by the UK Economics and Social Research Council and the European Commission. Appointments to advisory and governance boards include the UK BBSRC Science and Society Panel, and the Central Ethics Committee for Stem Cell Research, German Federal Government, Berlin.

Karakaya, Ahmet
University of Exeter, Department of Sociology, Philosophy and Anthropology, UK, and Medeniyet University, Faculty of Medicine, Department of History of Medicine and Ethics, Istanbul, Turkey

Ahmet Karakaya graduated from a Chemical Engineering Department at Yeditepe University, Istanbul, and continued his studies with an MA in Civilizational Studies at Fatih Sultan Mehmet University, Istanbul. His MA dissertation was on the ethical assessment of human embryonic stem cell research according to Turkish-Muslim scholars. He then worked for several years as Vice Director of Besikcizade Center for Medical Humanities in Istanbul. In January 2018 Ahmet began his studies for a PhD in Bioethics at Exeter University under the supervision of Christine Hauskeller. He is interested in the philosophical background of biomedical ethics, norms, and values in different Muslim bioethical debates, and ethical issues at the beginning and at the end of life.

MacGregor, Casimir
Sociology, School of Social Sciences, Monash University, Australia

Casimir MacGregor is a Research Fellow in the School of Social Science, Monash University. His research examines biopolitical infrastructures – how life, power, and social-material assemblages converge, intersect, and facilitate the flow of goods, people, or ideas, and allow for their exchange over space. He received his PhD in Social Anthropology from Macquarie University, which examined bioethical governance within the Australian human embryonic stem cell and cloning debate. His most recent research has examined the bioethics of human gene editing and informed consent within induced pluripotent stem cell (iPSC) research.

Manzeschke, Arne
Lutheran University of Applied Sciences, Nuremberg, Germany

Arne Manzeschke studied theology and philosophy at the universities of Munich, Tübingen, and Erlangen, received a pastoral's degree, and completed his doctoral thesis and habilitation in Erlangen. He is Professor of Anthropology and Ethics at the Lutheran University of Applied Sciences Nuremberg, Director of the Institute of Ethics and Anthropology in Health Care, President of Societas Ethica, European Society for research in ethics, and vice chairman of the Bavarian Ethics Commission on Preimplantation Genetic Diagnosis. Arne was a member of ForIPS – Bavarian Research Network Induced Pluripotent Stem Cells. He conducts ethical research in the field of bioethics and the ethics of technology, especially digitalization and the human-machine interaction.

Molnár-Gábor, Fruzsina
Ruprecht-Karls-University of Heidelberg and Heidelberg Academy of Sciences and Humanities, Germany

Fruzsina Molnár-Gábor (Dr. iur.) studied law in Budapest, Hungary, and in Heidelberg, Germany. She took her state examination in 2010 at the

Loránd Eötvös University, Budapest (*summa cum laude*). Until 2015 she was a Research Fellow at the Ruprecht-Karls-University and at the Max Planck Institute for Comparative Public Law and International Law in Heidelberg working on public international law, medical law, and bioethics. She completed her dissertation in 2015 in Heidelberg on the United Nations Education, Scientific and Cultural Organization's (UNESCO) work on the human genome and human rights (*summa cum laude*, prize of the Verwertungsgesellschaft Wort). Since 2015 she has been a group leader at the Heidelberg Academy of Sciences and Humanities on data protection and intellectual property law. Fruzsina is a lecturer at the Legal Faculty of the Ruprecht-Karls-University, Heidelberg.

Munsie, Megan
Centre for Stem Cell Systems, Department of Anatomy and Neuroscience, The University of Melbourne, Australia
   Professor Megan Munsie is Head of the Education, Ethics, Law and Community Awareness Unit at the Australian Research Council-funded Stem Cells Australia initiative. Megan is also Deputy Director of the Centre for Stem Cell Systems at the University of Melbourne, Australia, where she leads the Ethical, Legal and Social Implications Program. Megan is co-author of the recently published book *Stem Cell Tourism and the Political Economy of Hope* (Palgrave, 2017) (with Alan Petersen, Claire Tanner, Casimir MacGregor, and Jane Brophy) and the recipient of the highly esteemed International Society for Stem Cell Research 2018 Public Service Award.

Petersen, Alan
School of Social Sciences, Monash University, Australia
   Alan Petersen is Professor of Sociology in the School of Social Sciences at Monash University. His most recent books are *Hope in Health: The Socio-Politics of Optimism* (Palgrave, 2015) and *Stem Cell Tourism and the Political Economy of Hope* (Palgrave, 2017) (with Megan Munsie, Claire Tanner, Casimir MacGregor, and Jane Brophy).

Pichl, Anja
Department of Philosophy, Bielefeld University, Germany
   Anja Pichl is a research fellow at the interdisciplinary working group *Gentechnology Report* at the Berlin-Brandenburg Academy of Sciences (BBAW) and member of the research training group 2073 *Integrating Ethics and Epistemology of Scientific Research* at Bielefeld University, funded by the Deutsche Forschungsgesellschaft (DFG). Her PhD research in philosophy investigates stem cell research with regard to the scope of causal knowledge about stem cells and tensions between scientists' aims of understanding and intervention. She completed her MA in Philosophy at Ludwig-Maximilians-University Munich (LMU). Together with Arne Manzeschke, she organized the international summer school *Pluripotent stem cells: scientific practice, ethical, legal, and social commentary*, funded

by the German Ministry of Education and Research (BMBF), and worked on ethical issues within the Bavarian Research Network Induced Pluripotent Stem Cells (ForIPS), funded by the Bavarian State Ministry of Science and the Arts.

Prots, Iryna
Department of Stem Cell Biology, Friedrich-Alexander-University (FAU) Erlangen-Nuremberg

Iryna Prots is currently working as a group leader and performing her habilitation in the Department of Stem Cell Biology at the University of Erlangen-Nuremberg. She studied biology in Kiev, Ukraine, and then moved to Jena, Germany, within a student exchange programme, where she investigated the cytoskeleton in mammalian cells. She performed her doctoral thesis in clinical immunology at the University of Erlangen-Nuremberg, followed by postdoctoral studies in rheumatology and immunology at the University of Munich, before she joined her current laboratory at the University of Erlangen-Nuremberg. Her research focusses on neuroinflammatory mechanisms in human neurodegenerative disorders by establishing cellular model systems using human stem cells. She is a member of ForIPS – Bavarian Research Network Induced Pluripotent Stem Cells.

Rosemann, Achim
Department of Anthropology, Sociology and Philosophy, University of Exeter

Achim Rosemann is a Research Fellow at Egenis – the Centre for the Study of the Life Sciences (in the Department of Sociology, Philosophy and Anthropology) at the University of Exeter, UK. His research addresses the social, political, regulatory, and economic dimensions of life and health science research, with a regional focus on developments in China, East Asia, the UK, and the USA. He received his PhD in Social Anthropology from the University of Sussex, with a specialization in Science and Technology Studies and Medical Anthropology. He has previously worked at the University of Warwick (Sociology and Education Studies) and is also a Research Associate at the Centre for Bionetworking of the University of Sussex.

Sontag, Stephanie
Institute for Biomedical Engineering, RWTH Aachen University, Aachen, Germany

Stephanie Sontag is a research scientist at Taconic Biosciences GmbH in Cologne, Germany, where she works on genome engineering tools and on methods of constructing animal models for research. She studied biotechnology in Germany, Australia, and the USA, and obtained her PhD from RWTH Aachen University in 2017. In her dissertation she investigated blood cell development (particularly antigen presenting dendritic cells) from human pluripotent stem cells using CRISPR/Cas genome engineering.

Tanner, Claire
Centre for Stem Cell Systems, Department of Anatomy and Neuroscience, The University of Melbourne, Australia

Claire Tanner is a Research Fellow in the Ethical, Legal and Social Implications Program (ELSI) in the Centre for Stem Cell Systems at the University of Melbourne, Australia. Before joining the Centre for Stem Cell Systems, Claire was a Lecturer in Sociology at Monash University, where she was awarded her PhD in Gender Studies in 2011. Claire's research spans the sociology of science and health with a current focus on the social and ethical implications of new and emerging technologies. She is co-author of *Stem Cell Tourism and the Political Economy of Hope* (Palgrave, 2017) (with Alan Petersen, Megan Munsie, Casimir MacGregor, and Jane Brophy) and lead author of *Vanity: 21st Century Selves* (Palgrave, 2014) (with Jane Maree Maher and Suzanne Fraser).

Winkler, Jürgen
Department of Molecular Neurology, FAU Erlangen-Nuremberg

Jürgen Winkler is head of the Department of Molecular Neurology at the University Hospital Erlangen. He studied medicine in Freiburg and Strasbourg, performed his medical doctor thesis at the University of Freiburg and did his habilitation at the University of Regensburg. Jürgen is a specialist in neurology and successfully combined clinical practice and basic research during his career at the Universities of Düsseldorf, Würzburg, Regensburg, the Neurology Clinic Landshut, and the University of California in San Diego, USA. At the University Hospital Erlangen, Jürgen established a new department, where he is leading a group of laboratory and clinical researchers, which provides medical care to patients with movement disorders accompanied by the investigation of the pathological mechanisms of these diseases with the aim of developing new diagnostic and therapeutic options. He is the speaker of ForIPS – the Bavarian Research Network Induced Pluripotent Stem Cells.

Winner, Beate
Department of Stem Cell Biology, Friedrich-Alexander-University (FAU) Erlangen-Nuremberg

Beate Winner is the head of the Department of Stem Cell Biology at the University of Erlangen-Nuremberg. Following her studies of medicine at the Universities of Regensburg, Würzburg, and Toronto, she completed her residency and habilitation in neurology at the University of Regensburg. After a research period at the Salk Institute, La Jolla, USA, she started her own laboratory at the University of Erlangen-Nuremberg where she has focussed on modelling human neurodegenerative diseases using iPSC to investigate the molecular and cellular pathology of these diseases with the ultimate goal to develop new therapeutic strategies. She is a member of ForIPS – Bavarian Research Network Induced Pluripotent Stem Cells.

Zenke, Martin
Medical School and Helmholtz Institute for Biomedical Engineering RWTH Aachen University, Germany

Martin Zenke is chairman and Director of the Institute for Biomedical Engineering, Department of Cell Biology at Rheinisch Westfaelische Technische Hochschule Aachen University Medical School and Helmholtz Institute for Biomedical Engineering, RWTH Aachen University, Aachen, Germany. His research focusses on stem cells, stem cell engineering, and the application of stem cell research for disease modelling and drug screening. A particular emphasis is on pluripotent stem cells (embryonic stem cells and induced pluripotent stem cells), and hematopoietic stem cells and their differentiated progeny. Martin studied biochemistry and cell biology at Universities of Marburg and Heidelberg and at the German Cancer Research Center, DKFZ, Heidelberg. He worked at the University of Strasbourg, France, and at the European Molecular Biology Laboratories in Heidelberg, at the Institute for Molecular Pathology, IMP, Vienna and at the Max Delbrueck Center for Molecular Medicine, MDC, Berlin. Now he is Professor at RWTH Aachen University, and, since 2008, he has been a member of the Central Ethics Committee for Stem Cell Research, German Federal Government, Berlin.

3 20